QUERCY

Texts by Michèle Aué
and Antoine de la Viera
Translated by Alison Hebborn
and Barbara Davoust

MSM

SUMMARY

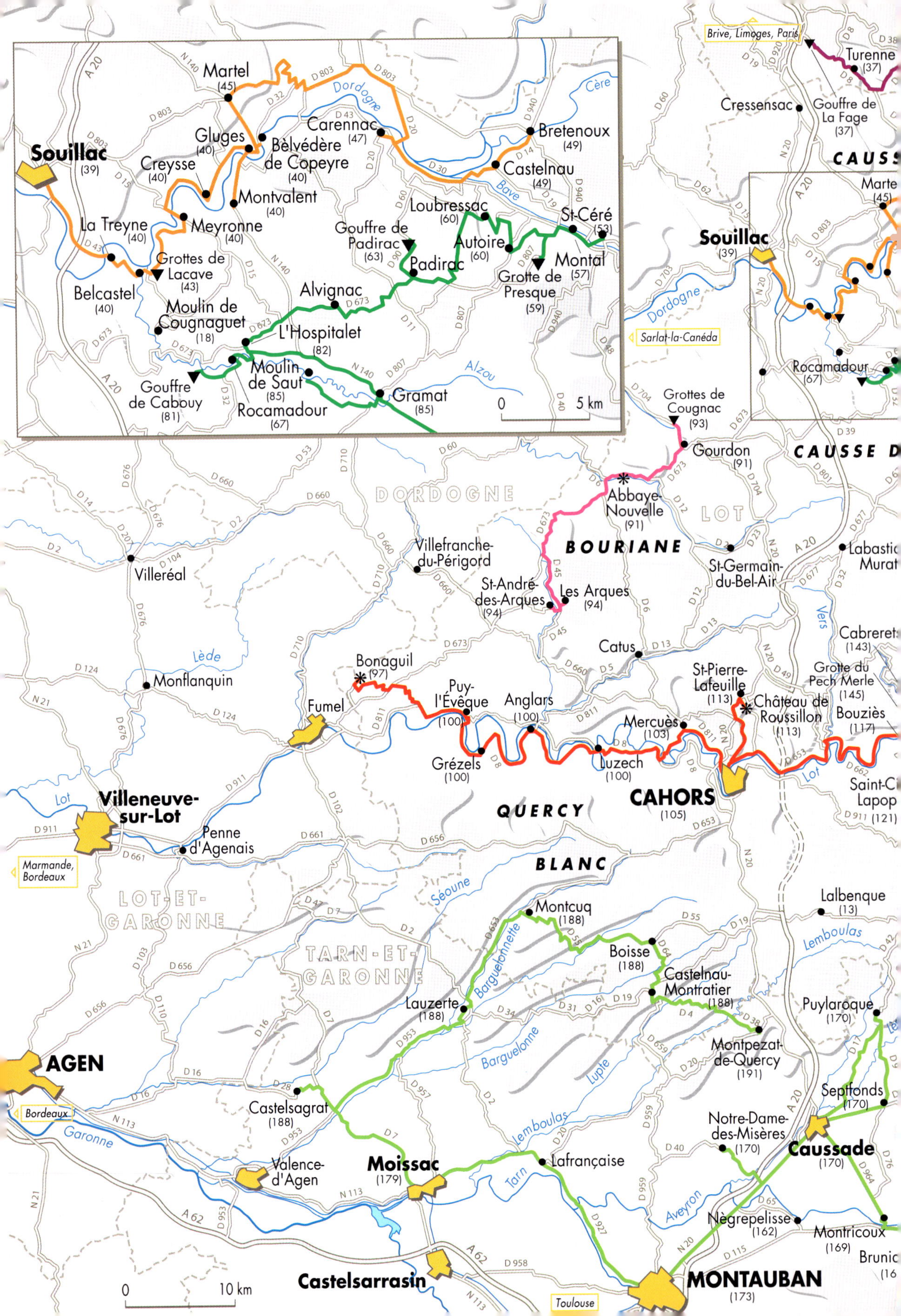
Souillac (39)
Martel (45)
Gluges (40)
Creysse (40)
Belvédère de Copeyre (40)
Carennac (47)
Bretenoux (49)
Castelnau (49)
Montvalent (40)
Meyronne (40)
La Treyne (40)
Grottes de Lacave (43)
Belcastel (40)
Moulin de Cougnaguet (18)
Gouffre de Cabouy (81)
Moulin de Saut (85)
Rocamadour (67)
L'Hospitalet (82)
Alvignac
Gramat (85)
Loubressac (60)
Gouffre de Padirac (63)
Padirac
Autoire (60)
St-Céré (53)
Montal (57)
Grotte de Presque (59)
Dordogne
Cère
Bave
Alzou
0 5 km
Brive, Limoges, Paris
Turenne (37)
Cressensac
Gouffre de La Fage (37)
CAUSSE
Sarlat-la-Canéda
Grottes de Cougnac (93)
Gourdon (91)
CAUSSE DE
Abbaye-Nouvelle (91)
DORDOGNE
LOT
BOURIANE
Villefranche-du-Périgord
St-Germain-du-Bel-Air
Labastide-Murat
Villeréal
St-André-des-Arques (94)
Les Arques (94)
Catus
Cabrerets (143)
Grotte du Pech Merle (145)
Lède
Monflanquin
Bonaguil (97)
Fumel
Puy-l'Évêque (100)
Anglars (100)
St-Pierre-Lafeuille (113)
Château de Roussillon (113)
Bouziès (117)
Vers
Mercuès (103)
Grézels (100)
Luzech (100)
CAHORS (105)
Saint-Cirq-Lapopie (121)
Lot
Villeneuve-sur-Lot
Penne d'Agenais
Marmande, Bordeaux
QUERCY
BLANC
LOT-ET-GARONNE
Séoune
Montcuq (188)
Lalbenque (13)
Boisse (188)
Castelnau-Montratier (188)
Lemboulas
TARN-ET-GARONNE
Barguelonnette
Lauzerte (188)
Puylaroque (170)
Barguelonne
Montpezat-de-Quercy (191)
Lupte
AGEN
Castelsagrat (188)
Septfonds (170)
Bordeaux
Garonne
Notre-Dame-des-Misères (170)
Caussade (170)
Valence-d'Agen
Moissac (179)
Lafrançaise
Tarn
Aveyron
Nègrepelisse (162)
Montricoux (169)
Bruniquel
Castelsarrasin
MONTAUBAN (173)
Toulouse
0 10 km

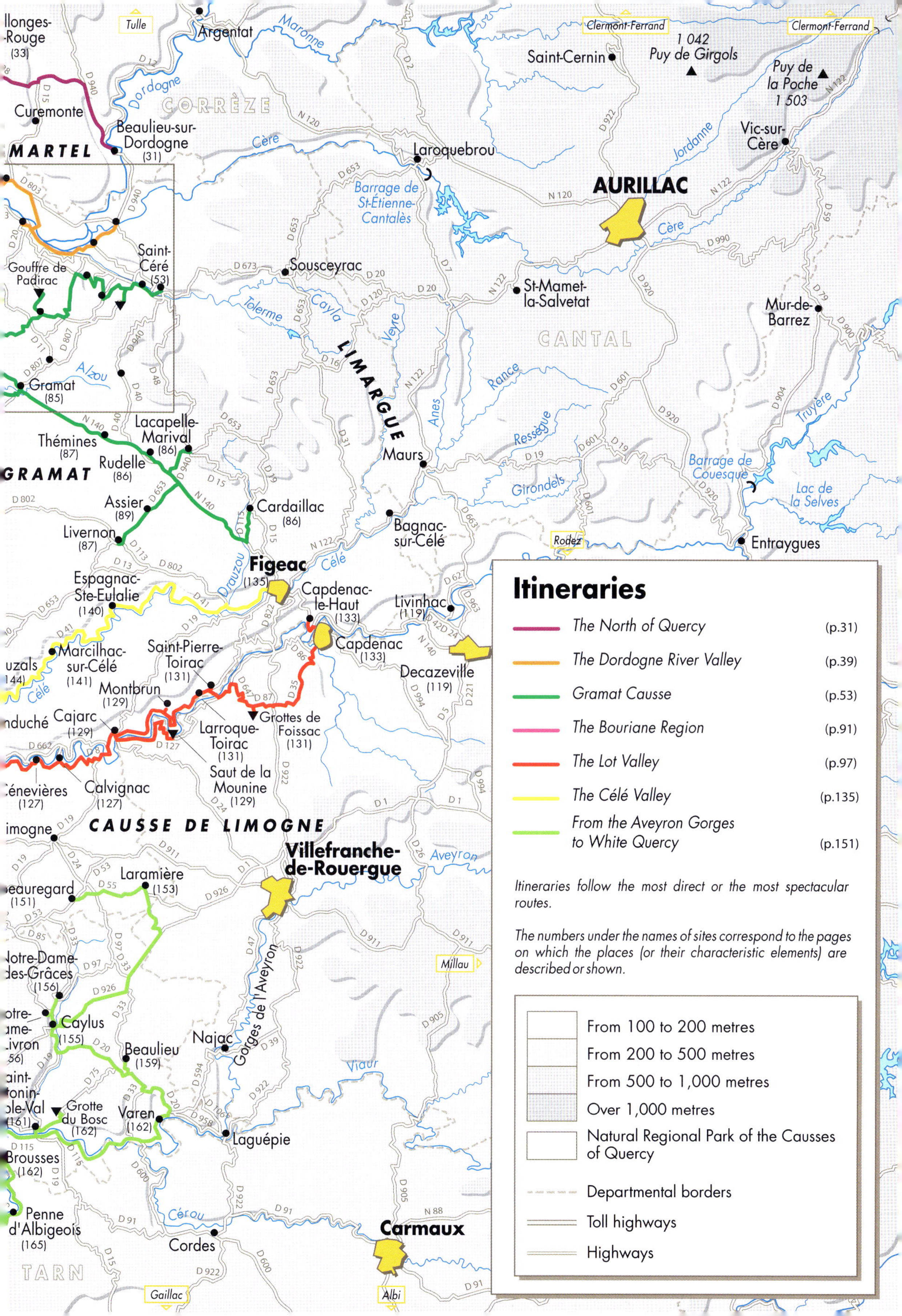

Tulle
Argentat
Maronne
Clermont-Ferrand
Clermont-Ferrand
Saint-Cernin
1 042
Puy de Girgols
Puy de
la Poche
1 503
Vic-sur-
Cère
Jordanne
AURILLAC
Cère
Laroquebrou
Barrage de
St-Étienne-
Cantalès
Dordogne
CORRÈZE
Curemonte
Beaulieu-sur-
Dordogne
(31)
MARTEL
Saint-
Céré
(53)
Gouffre de
Padirac
Alzou
Gramat
(85)
Thémines
(87)
Lacapelle-
Marival
(86)
Rudelle
(86)
GRAMAT
Assier
(89)
Livernon
(87)
Cardaillac
(86)
Sousceyrac
Tolerme
Cayla
Veyre
LIMARGUE
Anes
Rance
Ressègue
Girondels
Maurs
St-Mamet-
la-Salvetat
CANTAL
Mur-de-
Barrez
Truyère
Barrage de
Couesque
Lac de
la Selves
Entraygues
Rodez
Bagnac-
sur-Célé
Figeac
(135)
Célé
Drauzou
Espagnac-
Ste-Eulalie
(140)
Capdenac-
le-Haut
(133)
Livinhac
(119)
Capdenac
(133)
Decazeville
(119)
Marcilhac-
sur-Célé
(141)
Saint-Pierre-
Toirac
(131)
Montbrun
(129)
Cajarc
(129)
Larroque-
Toirac
(131)
Grottes de
Foissac
(131)
Calvignac
(127)
Saut de la
Mounine
(129)
CAUSSE DE LIMOGNE
Villefranche-
de-Rouergue
Aveyron
Laramière
(153)
Millau
Caylus
(155)
Beaulieu
(159)
Najac
Gorges de l'Aveyron
Viaur
Grotte
du Bosc
(162)
Varen
(162)
Laguépie
Brousses
(162)
Penne
d'Albigeois
(165)
Cérou
Cordes
Carmaux
TARN
Gaillac
Albi
Itineraries
The North of Quercy (p.31)
The Dordogne River Valley (p.39)
Gramat Causse (p.53)
The Bouriane Region (p.91)
The Lot Valley (p.97)
The Célé Valley (p.135)
From the Aveyron Gorges to White Quercy (p.151)
Itineraries follow the most direct or the most spectacular routes.
The numbers under the names of sites correspond to the pages on which the places (or their characteristic elements) are described or shown.
From 100 to 200 metres
From 200 to 500 metres
From 500 to 1,000 metres
Over 1,000 metres
Natural Regional Park of the Causses of Quercy
Departmental borders
Toll highways
Highways

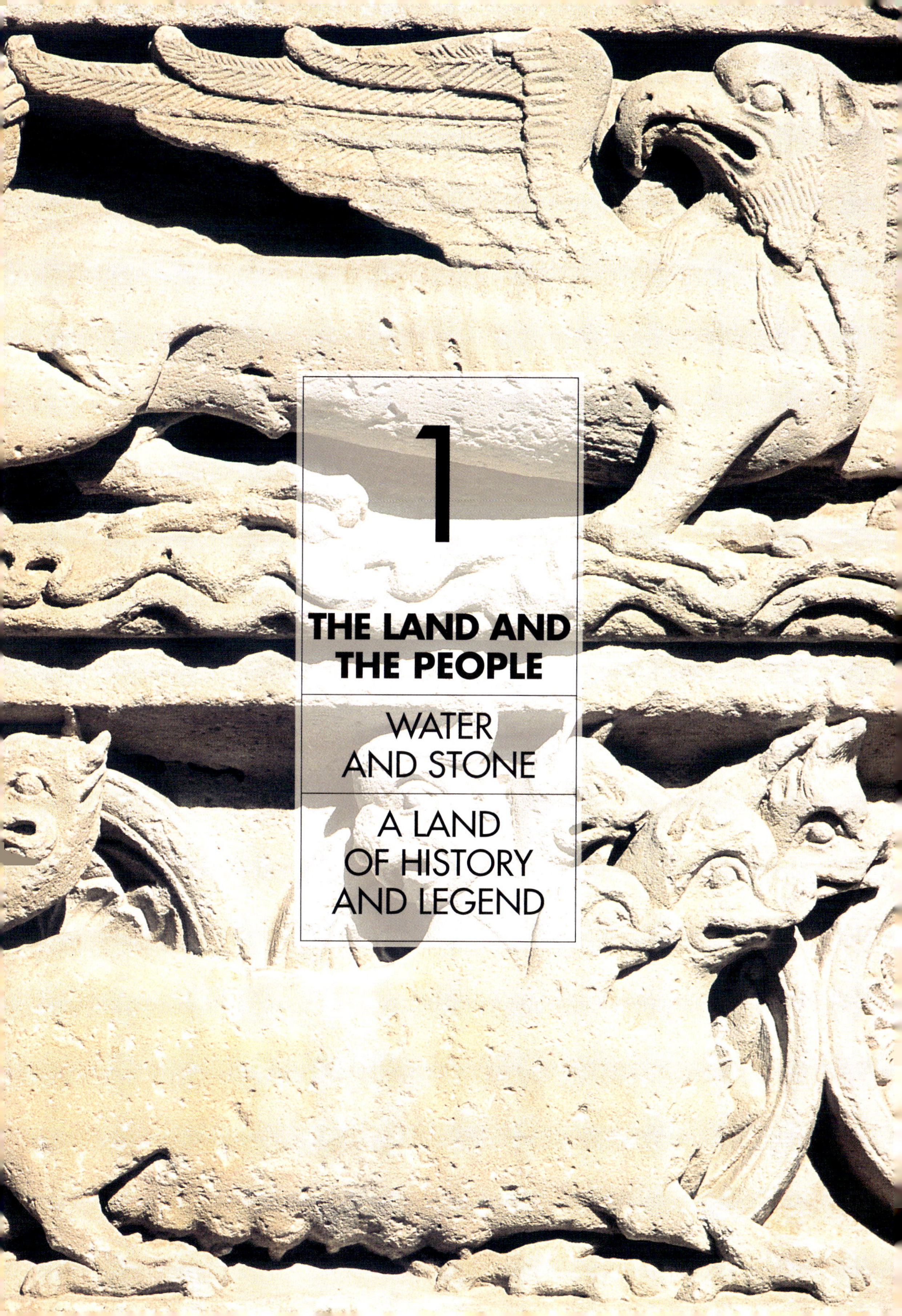

1

THE LAND AND THE PEOPLE

WATER AND STONE

A LAND OF HISTORY AND LEGEND

1

WATER AND STONE

Sheltered beneath the shallow seas of the Mesozoic era, great thicknesses of limestone sediment were slowly and silently laid down in secret. When at last the waters receded, they uncovered an area of limestone outcrop – the Quercy region and its limestone plateaus or causses. Although these plateaus can compare neither in size nor in altitude to those of the Grands Causses in the southeast Massif Central region, they do display all the typical features of a karstic landscape, which lends the area the essence of its distinctive personality. Crisscrossed by rivers sunk deep into the thick plateau, this varied region resembles some kind of huge jigsaw puzzle, its irregularly shaped pieces loosely interlocking, fitting a thousand fragile colours and subtle shades together into a single harmonious picture.

The region in the northernmost corner of Quercy, on the edge of the Martel Plateau, has a flavour and character all of its own. It boasts two places of particular interest: the fine medieval stronghold of Turenne, symbol of the power of the rich viscounts of the same name; and the little town of Collonges-la-Rouge – the pride of the region – radiant in shades of dark garnet red and bright crimson.

Pressed up against the complex contours of the Brive basin and the last spurs of the Limousin Plateau, the Martel Plateau stretches between the Vézère and Dordogne rivers. With some delightfully varied and untidy landscapes, the Martel Plateau presents much more than the bare uniformity that its rather flat name might suggest. The natural rolling contours of the plateau are highlighted by a haphazard

Collonges-la-Rouge: Ramade de Friac House

◄ Turenne: ruins of the feudal castle and Caesar's Tower

sprinkling of round, flat-bottomed hollows or dolines, known locally as *cloups*. These small fertile basins, filled with red limestone soil, bring the colours of the countryside to life and reveal, with the changing seasons, an ever richer palette of subtler hues. The Dordogne, threading its way between the Martel and Gramat plateaus, strings together the pearls that lie scattered along the valley, from Castelnau-Bretenoux, set slightly back on the Cère, through Carennac, Martel, Montvalent and so down to Souillac.

Bordered by the virtually parallel courses of the Dordogne to the north and the Lot and Célé to the south, and more extensive than the Martel Plateau, the Gramat Plateau provides a perfect example of a karstic landscape. Underlying rocks of exceptional permeability lend this area its general and characteristic barrenness. Rain – falling in abundance on the high plateaus – seeps, oozes and drips through a thousand cracks and crevices and disappears almost immediately into the complex labyrinth

Creysse: the church's chevet

A Heritage to Preserve

In the heart of the Lot department, the Natural Regional Park of the Causses of Quercy extends from a line through Belfort-du-Quercy – Laramière in the south, to the right bank of the Dordogne River in the north, forming an oblong territory consisting of the causses of Limogne and Gramat. On its 175,717 hectares lie 97 communes, in which live hardly more than 25,000 inhabitants. The density of this population, rather aging and constantly decreasing, is hence very low. It is therefore incapable of ensuring the ecological equilibrium and the preservation of its rural heritage. Houses stand abandoned, small roads are no longer maintained, and stone walls that were built up with such patience are slowly disintegrating. Land is left fallow. Both animal and plant species are threatened with disappearance. Many tourists come to visit, but this is just a seasonal business. Furthermore, it is concentrated in certain exceptional places, such as Rocamadour and Padirac. Tourists rarely venture to discover the quieter rural areas. The Natural Regional Park was therefore created in 1999 as an effort towards the restoration, protection, development and animation of the countryside – the property and pride of all who live in Quercy. The task ahead is enormous, but much has already been accomplished in a very short period of time. Country roads have been developed, tumbled-down walls have been rebuilt. Riverbeds have been restored, since the park is crossed by beautiful rivers, not only the Dordogne and Lot, but also small ones such as the Célé, the Vers, and the Alzou. The Park supports the action of the Quercy Phosphate Quarries Association (Phosphatières), that has opened to the public the mine at Cloup d'Aural and hopes to have the former mines registered as Unesco World Heritage Sites. The park's logo, a negative hand from Pech Merle Cave, is an invitation to discover the Quercy causses, but also a gesture trying to defend them.

Cougnaguet Mill ▶

that runs through the hidden layers of limestone. Here, unaided by seismic activity or tectonic upheaval, the physical and chemical action of water alone – eroding and sweeping away, dissolving and laying down deposits – has produced huge sculptured masterpieces of the most intricate workmanship.

Water filters down through potholes, carving out cavities. It swirls around the ceilings of subterranean galleries, hollowing them out into domes. It courses through mysterious circuits of sumps, disappearing through sinkholes into the bowels of the earth and is, seemingly, lost for ever. Then, chancing upon a bed of impermeable rock, it breaks loose from the stony confines of captivity and, with an explosion of crystal bubbles, gushes forth in a beautiful resurgence, like the one at Cabouy Chasm near Rocamadour, or as an abundant spring of clear cool water.

Sometimes the river abandons part of its subterranean course, leaving deserted caves to be furnished by the decorative talents of seeping rainwater, as it builds simple dripstone deposits into the architectural wonders of stalagmites, stalactites and columns. Every imaginable form of subterranean water erosion can be seen and admired in the shadowy depths of the Gouffre de Padirac, or Padirac Chasm – formed when the thin dome of rock that once protected its deep underground galleries collapsed, leaving a gaping hole.

But not all the water on the plateau disappears into the depths of the earth and, when a river remains on the surface, the relentless flow of its waters, together with the forces of freeze and thaw, combine with terrible efficiency to eat away at the sheer cliff faces with astonishing ease. Valleys become canyons as rivers cut downwards between high overshadowing walls. The Alzou and Ouysse valleys, where narrow ribbons of emerald green are surrounded by tall cliffs of solid limestone, were built like this. The sides of the twin valleys of the Lot and Célé rivers, from Figeac to Conduché, then the single channel opened out by

Black Diamonds

The chalky soils of the Quercy causses are particularly favourable for raising truffles. This mushroom has become the speciality of the area around Lalbenque, on the Limogne causse, and it is highly appreciated by food lovers. The truffle, known to botanists as *tuber melanosporum,* grows especially well among the rootlets of oak trees. As a parasite, it takes its nourishment from elements synthesized by the tree. It is picked as soon as the oak produces its first acorns. In order to pick these mushrooms, which grow under the soil, man uses animals with a highly developed sense of smell, like pigs, dogs, or even a special kind of fly, which, having picked up the smell, lays its eggs on the soil covering the truffles. For many years, truffles were picked in wild areas, but at the beginning of the 20th century, they were domesticated on soils favourable to their growth. The truffle grower plants oaks (holm or pubescent) or hazelnut bushes with mycorhiza. The tubercules of their roots carry the truffles' mycelium. After a period of anywhere from five to twenty years, a "burnt" zone may appear around the tree. The lack of grass here is a sign that the coveted mushrooms will soon appear. Nowadays, annual truffle production in the Lot department varies between three and ten tonnes and is sold on the markets of Lalbenque and Limogne, on each Tuesday of the winter season. The price varies, but in the past few years it has ranged between three hundred and seven hundred euros a kilogram.

Truffles

the Lot down to the borders of the Agen region, also feature abrupt drops, but with a less severe profile, as fragments of softer limestone accumulate at the base of the valley walls in gentle slopes, known locally as travers, or ribs. Meandering along below these slopes, the river becomes languid, tracing its course in gentle curves.

The near parallel lines of the great valleys of the Quercy region have ploughed the area into furrows of verdant farmland, alive with colour, where the richest treasures of the land are to be unearthed. Fruit-laden orchards, nestling alongside tender-leafed and gently rustling tobacco fields, are interspersed with sober rows of vines, whose fruit – the ruby-red wines of Cahors – bears that prestigious and well deserved mark of quality, the appellation d'origine contrôlée.

The Limogne Plateau, between the Lot and the Aveyron, is dryer, with rolling landscapes more typical of the South of France, scattered with dark patches of oak and juniper.

Falling away down gentle slopes towards the lower valley of the Garonne, is Quercy Blanc, or White Quercy, its lands divided into many long strips by the slender branches of its network of natural waterways, flowing to converge on the great river. It is, of course, the striking bands of tender chalk, lacerated by running streams, that lend this region its name.

An area of underlying sedimentary rock, Quercy is surrounded by a number of smaller, more diverse regions whose delightful jumble helps soften the contours of an otherwise severe landscape. To the east, the Limargue and Ségala regions slip into comfortable folds between Quercy and the Massif Central. To the west, Bouriane, wrapped in a fine cloak of forest, opens onto neighbouring Périgord. Although the old cries of battle are no longer to be heard, Bonaguil Castle, like a strong silent sentinel, watches still over this charmed land.

Finally, directly to the west, the Aveyron dawdles along in the last folds of the plateau until, as its waters are swelled by those of the Viaur

First-rate cooking

Quercy is a land of traditions and its cooking is issued from the resources of its native soil. Foies gras, pâté with truffles, magrets (breast meat) and confits (conserves) – all of these are the end products of a long process of selection, raising, and careful force-feeding of ducks and geese, and of local production, when not actually a cottage industry, in which the Quercy truffle sometimes finds its natural place. The cabécou, a little disk of goat cheese, is made by of goat farmers on the causses. It has the AOC (Appellation d'Origine Contrôlée) label. There is a technique (*couler*) of letting it get covered in a thick, downy skin that is tender under the tooth, then spreading it in gentle waves on a thick slice of bread, preferably with walnuts, since walnuts grow easily in the region and are delicious. Farmers' lamb from Quercy, raised on the causses, is fed exclusively with its mother's milk, hay and grains. Melons grow on the sunny hillsides. Even saffron, a rare spice produced from the pistil of the crocus, with its warm, ochre colour, prepared in Quercy, can add its rich flavour to a cuisine that already contains plenty. And of course, it is another local product, Cahors wine, that should be chosen as an accompaniment for some of these dishes and products, with which it harmonizes perfectly for the great pleasure of our taste buds.

Cabécous and walnuts

On the Geological Scale

Jean-André Poumarède was born in 1817 in Réalville, between Caussade and Montauban. He was a pharmacist, chemist and doctor. He was also an adventurer interested in everything. He travelled a lot, developed papyrin, an imitation parchment, but without patenting it, hence leaving others to make it and earn profits. He did not hesitate to volunteer to try the first diving-suit. He went off to exploit a silver mine in Mexico and was chased out by the civil war. Back home, in 1865 he stayed with his brother-in-law, who owned the domain of Cos, near Caylus. During a walk, Poumarède noticed that wheat was particularly vigorous in areas where the land was rich in certain nodules. Using his talents as a chemist, he analyzed them and found calcium phosphate, a natural fertilizer. He had immediately understood the value of his discovery for agriculture and paleontology, but died shortly afterwards, in 1869. The deposits were mined on a large scale in the entire area, both in open mines and underground. In 1886, the hundred and sixty-one Quercy quarries employed two thousand and forty workers. The phosphate was transported on *gabares*, flat-bottomed boats, along the Lot River to the canal lateral to the Garonne River, or by railway. But the deposits ran out in fifteen years. Production became marginal after the First World War, and stopped completely during World War 2. The phosphate mine at Cloup d'Aural, near Bach, retains the memory of this activity. A guided tour explains the complex formation of calcium phosphate. It has a long history of over sixty million years. The cavities of the former operation are rich in original vegetation and many fossils.

and the Bonnette, it gathers force to carve out the beautiful steep-sided valley that aptly bears the name of the Aveyron Gorges.
This secret and isolated land was well suited to the lifestyles of the various communities of monks who came here in the Middle Ages in search of the kind of protected site that could afford them perfect peace and tranquillity. Beaulieu en Rouergue, like all Cistercian abbeys, is the very image of the perfection looked for by these hermits and men of God.
From Caylus, perched on its rock above the Bonnette, to Saint-Antonin-Noble-Val, Penne, Bruniquel and Montricoux on the Aveyron, the villages cling to their steep river banks, watching over all who come and go here.
Entering at last into lands of a less complex geography, the rivers of Quercy flow together, weaving a regular web. As the Aveyron swells the Tarn, and the Tarn flows into the Garonne, the valley widens, preparing the way for the opulence of the Agen region, with Quercy's pride and joy – Montauban and Moissac.

Beaulieu en Rouergue Abbey: the Easter Lamb

◄ The chasm of a phosphate mine at Cloup d'Aural near Bach

Country Architecture

The art of building has developed over the centuries ever since pre-historical times. But few regions have retained the memory of this art like Quercy. Quercy is a land of stone. The causse has always supplied it in large quantities: blocks of golden or gray limestone, easily carved, joined with lime, or simply piled up in dry-stone walls. From the low walls surrounding fields, to farmhouses with staircases leading to the upper living floor and their turret, to roofs made of large flat stones called lauzes, stone is always present. With its often barren soil, Quercy has forced man to improve the land. To compensate for not being able to leave land fallow for long periods, men began to fertilize soil with colombine, or pigeon guano. And in order to harvest enough of this fertilizer, during the 17th and 18th centuries, they built dovecotes, before turning to supplies of local phosphate deposits. These dovecotes are all very different. Some stand high on man-made blocks, others perch on stone piles and shelter under roofs with a central pinnacle. Shelters for shepherds, small barns called *gariottes,* or sometimes *caselles,* are secular witnesses of an architecture developed from necessity. Generally cylindrical, covered with a roof with a rounded summit, built of flat-stones, they form a perfect complement to the causse landscape. And Quercy's rivers, often hemmed in, have provided water power for numerous mills, such as those of Saut and Tournefeuille on the Alzou River, or the fortified mill at Cougnaguet on the Ouysse River, a Cistercian barn from the 14th century.

A *gariotte* near
Saint-Antonin-Noble-Val

Old house
in Loubressac

A dovecote
near Caylus

2

A LAND OF HISTORY AND LEGEND

For thousands of years, the limestone plateau has been dissolved, undermined and eaten away by the continuous action of water on stone, carving out extraordinary underground networks of caves, caverns and chasms. In these places of exceptional natural beauty, our distant ancestors found welcome shelters and hiding places. First Neanderthal man, then our closer cousin, Cro-Magnon man or *Homo sapiens sapiens*, explored these stony plateaus, discovering their deep galleries etched out of rock. Bouffia Bonneval, near La Chapelle-aux-Saints, the Cougnac caves, the beautiful, rich Pech Merle cave, as well as those in Foissac, have protected the fragile traces of their passage until today.
The strange prehistoric stone monuments, or megaliths, that litter the region, date from a

Belinac Menhir near Livernon

La Chapelle-aux-Saints

In August 1908, the priests Amédée and Jean Bouyssonie, with their young brother Paul, explored the Bouffia Bonneval, a natural cave in the valley of the Sourdoire River, a tributary of the Dordogne, near the hamlet of La Chapelle-aux-Saints (the Saints' Chapel) in Corrèze. They found a silex point, a bison horn, and bison and reindeer vertebrae. On the 3rd, they discovered a nearly complete skeleton of a Neanderthal man, *Homo neandertalensis*, with his stocky body, and his large head with its receding forehead resting on a continuous, suborbital fold. The man had been buried by his kin in a sleeping position. Measuring about 1.65 metres, he was rather short compared to male standards of the group. He was well over 50 years old, whereas life expectancy was rarely more than 40. Remains of animals found on the site attest to the Arctic-like climate of the time when he lived, placing it between 63,000 and 40,000 years ago. The abundance of stone tools (axes, scrapers, saws, drills, points, blades) found, of which over a thousand pieces are kept at the Labenche Museum in Brive, shows that the cave was a workshop where silex and quartz were cut and polished, or else, a dwelling-place where these activities were carried out. In the museum opened near the site in 1996, the burial site was reconstituted. Documents explain the history of its discovery. *Homo neandertalensis* and *Homo sapiens sapiens* are compared and placed within the context of evolution. There is a description of the life and environment of the ancient old man found in La-Chapelle-aux-Saints.

later age. Menhirs, or standing stones, rising abruptly among juniper and oak, and heavy dolmen tables, with rough arches that open silently onto mysterious burial places, are so abundant in Quercy, that the region has one of the highest concentrations of Neolithic remains to be found anywhere in France.
As prehistory emerged into the age of history, the written records that were kept to recount and explain, sometimes created confusion and mystery. The Cadurques were a Gallic people who did not take kindly to systematic conquest by Roman legions. In an attempt at organised resistance, two thousand Cadurques barricaded themselves into the famous fortress town of Uxellodunum, so eloquently described by Julius Caesar himself in his writings on the Gallic wars. Caesar, cunning and implacable, set out to demonstrate the greatness of his military genius. The Roman armies laid siege to the little stronghold but the Cadurques held firm. Caesar, becoming impatient, ordered his legionnaires, who were used to exchanging their role of soldier for that of navvy, to alter the course of the spring that supplied his enemies with water. Thirst was a powerful enemy,and the Cadurques were forced to abandon their struggle. Caesar, setting a cruel example as a deterrent to any other would-be rebel Gauls, sentenced them to have their right hand cut off. Oddly enough, although all the human, technical and strategic details of this episode are recorded, it is impossible to locate the stronghold of Uxellodunum with any degree of certainty. Perhaps the ancient authors thought it unnecessary to describe the position of such a well-known place, because they left no indication of the site's geographical location. As a result, there is a local saying that "the fight for Uxellodunum is not over yet". Many sites, including Capdenac-le-Haut, Puy d'Issolud, Luzech and Murcens, claim the right to bear this legendary name. But, until archeological research unearths enough conclusive evidence to support one or other of their cases, the matter cannot be settled.

The Roman peace, with its mixture of humiliating defeat and peaceful prosperity, gave birth to an opulent new hybrid civilization – the Gallo-Romans.
Invaded and occupied by the Visigoths and then the Franks, Quercy moved fully into the Middle Ages. Towards the year one thousand, Christianity was seized with great mystical fervor. As a Cluny monk, Raoul Glaber, wrote at the time, "One would have said that the whole world, of a common accord, had thrown off its ancient rags to cover itself with a white robe of churches." It was the dawn of Romanesque art that, for two centuries, was to produce in Europe masterpieces of sculpture that can be seen in Quercy on the tympanums of Moissac, Carennac, Beaulieu-sur-Dordogne or Collonges-la-Rouge. It was this same fervor that would drain pilgrims towards the great pilgrimage sites: Compostela, Saint-Gilles, but also Rocamadour, splendid and hieratic on the side of its sheer cliff, welcoming, to the vast emptiness of the causse, those

Moissac, St. Peter's Abbey Church: Abraham

Souillac, St. Mary's Abbey Church: Romanesque sculpture of the Prophet Isaiah ▶

walking for God and coming to pray at the tomb of Saint Amadour.

Faith alone, however, was insufficient defense against the treacherous twists and turns of historical fortune, and citadels, castles, towers and fortifications soon sprang up along the heights. From the most formidable sites of Bonaguil, Bruniquel or Castelnau, to the more modest but still very impressive Larroque-Toirac or Cabrerets, Quercy is, today, littered with their remains.

But one also had to live, and also, to accumulate wealth. Trade became an essential preoccupation, and in order to protect its riches, a network of walled towns, bastides, sprang up all over Quercy at the beginning of the 14th century. From the major town of Montauban to little Rudelle, towns armoured with ramparts were created from scratch by civil or church authorities anxious to defend their wealth. Many simple villages, like Curemonte, Loubressac, and of course, Saint-Cirq-Lapopie, found relative security in their loca-

Loubressac: 15th and 16th century church

Saint-Cirq-Lapopie: The Rocamadour Gate and the site

tion at the tops of cliffs or hills. These were troubled times. The Albigensian Crusade starting in the 13th century, the Hundred Years War in the 14th and 15th, the Wars of Religion in the 16th century, without counting permanent peasant uprisings, ravaged Quercy, sparing neither man nor his works.

Like many regions of France, Quercy was greatly influenced by the art of the Italian Renaissance. This influence is brilliantly illustrated in the castle and church at Assier, on the Gramat causse, as well as Montal Castle near Saint Céré.

Then in the era of administrative division, heir of the 1793 Convention, two departments were created – Lot and Tarn-et-Garonne – splitting in two the beautiful province, birthplace of Murat, one of Napoleon's great Marshals. In the mid-19th century, Quercy's isolation ended with the arrival of the railways, but it was nonetheless able to retain its name and its soul and strength, in the heart of its harsh and grandiose landscapes.

Assier, the castle: doorway

A Child of La Bastide

Joachim Murat was born on March 25, 1767, the twelfth child of an innkeeper in La Bastide, a village in the south of Gramat causse. His family wanted him to become a man of the Church. In 1785, he entered the Lazarist seminary in Toulouse. But the young man ran up debts and, to escape his creditors, joined the army in 1787. He was thrown out for indiscipline. In 1791, Murat joined Louis XVI's constitutional guards. On 13 Vendémiaire (1st month of the Republican year), he was noticed by Napoleon, who made Murat his aide-de-camp. From then on, this Quercy native's destiny was linked to that of the brilliant Corsican he accompanied to Italy and Egypt. His behaviour at the second battle of Aboukir on July 25, 1799 led to his promotion as a division general. The day after 18 Brumaire, Murat led the grenadiers who chased out the deputies of the Council of Five Hundred. A grateful Bonaparte entrusted him with the command of the consular guard and the hand of his sister, Caroline. In 1804, Murat became a Marshal of the Empire. It was he who, in Madrid on May 2, 1808, put down the popular uprising in a bloodbath. In the same year, he became the King of Naples, then of the Two Sicilies. He participated in the Russian campaign in 1812, and led the retreat of the Grande Armée. Starting in 1813, he began to scheme with Napoleon's enemies but, knowing nevertheless "who made him king", was at his side in Dresden and Leipzig. In 1814, Murat joined with Austria and England, who preferred to see a Bourbon on the throne of the Kingdom of Naples. Murat felt betrayed. When Napoleon returned from the island of Elba, he tried in vain to regain his place. On October 8, 1815, Murat landed at Pizzo, in Calabria, to reconquer his kingdom. He was captured and executed five days later. Today, La Bastide has added the name of its most famous child to its place names.

Bioule, the castle, wall paintings in the Salle des Preux: Caesar ▶

2

ITINERARIES

CAUSSES AND RIVERS

1

2

1

THE NORTH OF QUERCY

Beaulieu-sur-Dordogne

Beaulieu-sur-Dordogne developed around an abbey founded in about 840 by the archbishop of Bourges, Raoul de Turenne, on this *bellus locus*, lands of the viscountcy held as a fief by his family. The original community was composed of monks who had come from Solignac in Limousin. In 1095 the abbey came under the Cluny rules and thereafter entered a period of prosperity and development. Between the end of the 11th century and the first half of the 12th, the abbey built a new abbey church, named for Saint Peter. During the 14th century, a second belfry, resembling a defence tower, was raised at the south-western corner. During the Wars of Religion, the abbey was pillaged and its community scattered. In 1663, monks from the Saint-Maur order arrived and life continued until the Revolution. The abbey buildings were then demolished. With its chapter house converted into a sacristy, only the abbey church was spared and turned into a parish church. The southern doorway of St. Peter's church, protected by a porch, is a remarkable example of Romanesque art. The tympanum was sculpted towards 1130, developing the theme of the Return of the Lord at the end of the world. In front of two angels holding the cross, another carrying the nails of the Passion, and a fourth with the crown of thorns, Christ holding his arms open is surrounded by two other angels who are blowing their trumpets. The apostles, as well as a prophet symbolizing the Old Testament, are present. On either side, groups of figures represent humanity, while the dead resuscitate and raise the lids of their tombs. A double lintel is illustrated: on the upper one, monsters are devouring the damned; on the lower one, the mortal sins are symbolically depicted. The pier presents figures carved on three sides. On the western side, a man is giving a leg up to another who, like a telamon, seems to be holding up the entablature that the lintel is resting on. In the south, a very elongated man alone appears to be taking part in the same task. On the eastern side, an old man with a large beard and a drooping moustache, seems to be stuck between two wavy columns. At their feet, Peter and Paul welcome the faithful. The sculptures on the porch walls have been damaged. On the left, lions lick Daniel's feet in the lion-pit, and on the right, the three temptations of Christ by the Devil are shown. Inside the church, the layout of the sanctuary is that of a pilgrimage church, with a nave, a barrel vault, flanked by two aisles with groined vaults, a transept and a chancel with two bays, encircled by a deambulatory. In the axial chapel of the latter, there is a reredos showing episodes in the life of the Virgin on its three panels: the Assumption, in the presence of the apostles, and the Coronation. A strongbox, located in the northern arm of the transept, contains the treasure consisting of a 13th century reliquary of champlevé enamel, a 10th century Byzantine reliquary, two reliquary arms from the 13th and 14th centuries, and other objects, including a magnificent 12th century statue of the Beaulieu Virgin. This wooden statue is covered with slashed silver leaf in which filigree, precious stones, and a cameo are encrusted. The Virgin is seated and holding, with her left arm, the Infant sitting in her lap and holding the Book, and in her right hand, holding what appears to be the mystical Almond, the symbol of the virginity of the Mother of Christ.

Collonges-la-Rouge

For those who do not know Collonges-la-Rouge ("the Red"), a careful study of a photograph might lead to the following comment: was the bright colour of the walls not slightly intensified by some technical process?
Collonges is called "the Red" and it is not an exaggeration!
The stone used – a superb sandstone with scarlet nuances – makes it an exceptional place. The range of colours that cover the town is extremely wide. The crimson of the building stones sets off the blue-gray of the roofs with their shiny reflections and stands out against the green background.
Collonges-la-Rouge is a former medieval town, a stopover for pilgrims on their way to Compostela leaving the *via Lemovicensis* to join the *via Podiensis*. On their way, they took advantage of the change to venerate the body of Saint Amadour at Rocamadour. The town's founding around a Benedictine priory depending on Charroux Abbey dates from the 8th century. A vassal of the viscountcy of Turenne, the town was enfranchised and hence obtained its communal freedom in the 13th century.
Its site was appreciated by civil or military officers of the Turenne viscountcy, who, bit by bit, created a small town with substantial houses, and with harmonious and graceful manor houses.
Strolling through Collonges-la-Rouge is enchanting. Footsteps resound on the shining paving stones of the old streets, filled with doorways with barreled arcades, lined with walls covered with Virginia creeper and wisteria. All the houses in Collonges-la-Rouge deserve a mention, given the charm of their towers, turrets, pepper-pot roofs and solid staircases, such as those we can admire along Rue de la Barrière. The Siren house, with its flat-stone roof opens onto the street through an arched porch. A strange little sculpture

Towards the Field of Stars

James the Greater was one of the twelve apostles, and brother of John. Martyred in 44 CE on the orders of Herod Agrippa 1st, legend says his body was taken away by his disciples, placed in a boat which, "guided by the hand of the Lord", landed on the Galician coast at the north-west of the Iberian peninsula. The saint's tomb was found in the years 820-830 by Pelayo, a hermit, who saw lights near Iria Flavia. The King of Asturia, Alfonso the Chaste, came there on a pilgrimage in 834, and had the first sanctuary built to shelter the apostle's tomb, at Compostela (from *campus stellae*, field of the star). Saint James, *matamoros*, then became the patron of the *Reconquista* and intervened in the legendary battle of Clavijo, near Logroño in 844. The pilgrimage developed. The first French pilgrim was the Bishop of Puy, Godescalc, in 950. The development of the pilgrimage was barely slowed by the raid of the Moorish chief Al Mansur in 997, who had the Compostela church destroyed. The religious and military orders, like those of Cluny, founded hospices along the paths taken by pilgrims coming from all of Europe. The four main paths were described by the Cluny monk, Aymeric Picaud, in his *Liber peregrinationis*, written in about 1130. *Via Tolosana, via Podensis, via Lemovicensis* and *via Turonensis* "are all one" at Puente la Reina to form the *camino Francés*.

Saint James

◄ Collonges-la-Rouge: St. Peter's Church

gives the house its name. On one of the facade's cornerstones, a mermaid combs her hair, holding a mirror in her other hand. Coquettish and child-like, she gives off a very particular sense of seduction. In the same street, not far from the Plate (flat) Gate, the Ramade de Friac House has a fine facade divided by two solid, cylindrical towers topped with pepper-pot roofs.
The market hall, tucked in near the church, is covered with handsome beams and contains the well-preserved remains of an old oven. The small Penitents' chapel from the 13th century is a sober masterpiece. Its wall-belfry prolongs a plain facade, and its massive buttresses fill out this tiny building. The coloured outlines of the church's steeple and towers stand out against the background of the blue sky. Built during the 11th and 12th centuries, it was subsequently fortified and flanked by a keep with a defence room and a covered way. The steeple, in pure Limousin Romanesque style, consists of several storeys fitted together, from its square base with its two rows of bays edged with fine columns, to the perfect geometry of its final octagonal storeys. The contrast of the white limestone used for the tympanum with the red surrounding walls is striking! At the top level, a standing Christ seems to rise gently towards Heaven, held up by two kneeling angels. Two other angels, with ample gestures, join in this Ascension. At the lower level, the Virgin and the apostles are carefully distributed under the arcades of celestial Jerusalem.
But Collonges-la-Rouge also has its castles, as well as small mansions that look rather like castles, called *castels*. Examples are Benge, close to the Plate Gate, with its round tower spiked at mid-height by the remains of old corbels; that of Vassinhac where medieval and Renaissance influences join, the residence of Gédéon, Lord of Collonges and governor of Turenne viscountcy; finally, that of Maussac with its bartizan hanging on at the top of its tower.

Collonges-la-Rouge: Maussac Castel

Collonges-la-Rouge: Ramade Mansion ▶

Turenne

Starting in the 10th century, the lords of Turenne won their independence from the kings of France. The viscounts of Turenne minted their own money, created nobles and raised taxes. Their domains, which in the 15th century included Upper Quercy, Lower Limousin and Sarladais, were a haven of prosperity. The privilege of not having to bear the fiscal levies weighing on the lower classes of the kingdom was so exceptional that the saying "happy as a *vicomtin*" became proverbial. But in 1738, Louis XIV bought the overly favoured and autonomous viscountcy for over four million *livres* from the ninth viscount of the lineage of the Tour d'Auvergne, who held the fief of Turenne. Taxes, levies and tribute of all kinds rained down on the province. The remains of Turenne castle still look proud even though the main part of the fortress was dismantled on the orders of Louis XV. The Clock (Horloge) Tower to the south, the former 13th century keep, and Caesar's Tower in the north, from the 11th century, rise at either end of the mound on which the castle stood. A wall runs along the edge of the slope and borders a pretty garden planted with dark boxwood. The ancient houses of the upper town, crossed by rue Droite, hang on to the side of the hill, while the lower town, the former suburbs, has retained its fine houses from the 15th, 16th and 17th centuries, such as Vachon House, former residence of the consuls of Turenne, opening onto the Place de la Halle (Market Square), the Ancien Chapitre House, located in rue Joseph-Rouveyrol, leading to the 16th and 17th century church. Between it and the castle hill, stands the Tournadour House, former salt storehouse, Sénéchal House, and the Capucin Chapel from the 17th century. North-west of the town, La Fage Gouffre (chasm) is the result of the collapse of the rock vault. Strangely resonant concretions line the various chambers.

"Well May You Tremble!"

Whether fact or fiction, historical quotes seem to work their way down through the centuries with surprising consistency. So it is that Henri II de La Tour d'Auvergne, Marshal General of Louis XIV's royal armies, is said to have rebuked himself: "Well may you tremble, my carcass! If you knew where we were going, you would tremble even more!". This brave soldier, from the famous La Tour d'Auvergne family, of the dynasty of the Viscounts of Turenne, fought and won the battles that, in the 17th century, were to enable the French monarchy to determine and defend the frontiers of France. Having won countless honours on the fields of battle, he met his untimely end at Sasbach, on July 27th 1675, struck down in full glory by a well aimed canon ball. He was accorded the ultimate, albeit posthumous, distinction of being buried among the country's most powerful dead, at Saint-Denis, and later in the Invalides. He is still remembered, in French school books as the "Great Turenne".

Henri II de La Tour d'Auvergne, Viscount of Turenne

2

THE DORDOGNE RIVER VALLEY

Souillac

Just one of the many little towns dotted along the Dordogne valley, Souillac is built on what used to be a wild boar wallow, known as a *souilh*. The task of draining this marshland site was first undertaken by the patient monks of the local abbey, thought to have been founded, in the 7th century, by good King Dagobert's dear Saint Eligius, and later attached to the Benedictine monastery at Aurillac.
Today, all that remains of Saint Martin's Parish Church is its solid but disfigured belfry. Saint Mary's Abbey Church, however, dating from the 12th century still bears witness to the past splendour of what was once an important religious centre. Built in a typically Byzantine style, the church has a fine chevet with white walls pierced by small regular openings and surrounded by apsidioles with blind arcatures, a central dome crowned with an openwork pinnacle, and wide-spanned roofs of a delicate lacy appearance, and is one of the finest examples of Romanesque architecture in the whole of southwest France. The line of its domes set carefully over the nave and the exhilarating contours of its vaulted roof lend the church a lightness and a purity of form that are quite remarkable. The richness of its decor contrasts with this harmony but by no means spoils it. The Romanesque sculptures that decorated the facade in the 12th century have been carefully preserved and now adorn the nave. Works of the Virgin Mary saving Theophilus from eternal damnation, of famished monsters with cruel claws, and of a particularly elegant Prophet Isaiah, stepping gracefully in the flowing folds of his lace-trimmed robe, represent some of the finest examples of the Toulouse school of Romanesque sculpture.

The Automaton Museum

Souillac's Automaton Museum, close by the abbey church, owes its fame and fortune entirely to the enthusiasm and astoundingly creative ingenuity of successive generations of the Roullet-Decamps family. In the middle of the 19th century, at a time when the miniaturisation of mechanisms and automated systems became a highly sought after skill, these inventive artists combined an unusual degree of manual dexterity with a certain good taste to imagine and create decorative mechanical objects. The Souillac museum's collection has more than a thousand pieces, ranging from pretty nodding faces with thoughtful expressions, to the graceful leaps and bounds of little dancers in their tutus, not forgetting the electrifying energy of a furiously rhythmic jazz band.

The clown on the moon

The Course of the Dordogne River in Quercy

The Dordogne River enters the Lot department north of Bretenoux, after having crossed Beaulieu-sur-Dordogne. The river then starts its winding course as far as Souillac, dividing the causses of Gramat in the south and Martel in the north. After the waters of the Céré River join it, the Dordogne divides into two reaches that delimit an oblong strip of land, Escoulans Island. Another island named Calypso, dear to Fénelon, has formed down-river, at the foot of the village of Carennac. South of Martel, the river draws a wide loop at the foot of an amphitheatre of high cliffs – Montvalent cirque. Across the river, beside the village of Gluges, by a semi-trogloditic 12th century church, decorated with Romanesque modillions, Copeyre belvedere provides a fine viewpoint over the river and the cirque. Creysse stands on another meander. Its Romanesque church, with its two, circular, attached apses, is the former chapel of a castle belonging to the Turenne viscountcy. On his way through Creysse, Saint Louis allegedly donated one of the Holy Thorns of Christ's crown to the church. South of Meyronne, former residence of the bishops of Tulle, the Ouysse River, coming from Gramat causse, flows into the Dordogne, near the Lacave caves. A rocky spur overlooks the confluence. It is topped by Belcastel castle, which has retained its chapel and a medieval main building. Further downstream, before the course of the Dordogne straightens out towards the north and Souillac, La Treyne castle overlooks the left bank of the river. Destroyed during the Wars of Religion, it was restored under Louis XIII and remained the property of the Cardaillac family for over two centuries.

Montvalent Cirque
seen from Copeyre
belvedere

La Treyne Castle

Belcastel Castle

◄ Lacave Caves: the Minaret and Gours

Lacave Caves

Among the cliffs that border the valley of the Dordogne, where natural erosion of the karstic landscape has carved out a fascinating hollow relief, Lacave Caves were discovered by Armand de Viré in 1902. An enormous cavity, more than twenty metres high, that was used as a place of shelter during the Paleolithic Age, continues along a gallery into the dark depths of the cliff, and leads to the Hall of Chaos, thrown into confusion long ago when earth tremors destroyed the cave's ceiling.
A series of chambers with unusual dripstone formations is strung out along the length of a beautiful gallery. Stalagmites and stalactites make fabulous and sometimes eerie shapes, such as the mythical *Tarasque* monster, watching, with an almost unnerving immobility, for the slightest shimmer in the clear reflective waters below. Such a wide variety of forms are molded and engraved so delicately onto the rock that it would almost seem strange not to be able to discern among them the shapes of *candles, chandeliers, organs,* a *palm tree* and an *elephant*. Beyond this profusion is the chamber of Our Lady of Lacave, where time and nature have modeled the calcite into the graceful outline of a *Madonna and Child*.
From the Chaos, another gallery opens out like a huge bubble into the high vaulted ceiling of the Great Dome, and leads on to the lacy *Spider's Legs* column, formed by the gentle oozing of thousands of drops of water. In the chamber of the Great Lake, the motionless mirror of water creates the perfectly symmetric illusion of an underwater town of fabulously decorated pinnacles, towers and minarets. But this fairy-tale kingdom is at its most enchanting in the last chamber, cleverly lit and fittingly named the Hall of Marvels, where bizarre forms – aptly called eccentrics – are built to unknown rules that seem to defy the law of gravity, and where imagination and exuberance appear to be the order of the day.

Before History

The objective of Préhistologia, a prehistorical theme park located between Rocamadour and Lacave, is to retrace the various stages of Prehistory. Along a path that explores acres of woods, life-size reconstructions of dinosaurs will certainly impress the visitor. It was not for nothing that the English paleontologist, Richard Owen (1804-1892), named them *dinosauria,* from the Greek words *deinos,* terrible, and *sauros,* lizard. These "terrible lizards" appeared on our planet 200 million years ago and disappeared, for reasons which are still unexplained, 60 million years ago. What kind of event could have wiped out the Stegosauruses, Styracosauruses, Triceratops, Brachiosauruses, Velociraptors and other Pteranodons? The appearance of man took place well after the disappearance of the dinosaurs – only about 4 million years ago. Man appeared and evolved. *Australopithecus* was followed by *Homo habilis,* then *Homo erectus,* finally *Homo neandertalensis* and *Homo sapiens sapiens,* similar to us. All were hunters and gatherers. Their prey consisted of mammoths, bears or reindeer, depending on the period. *Homo erectus* discovered fire 500,000 years ago and learned to master it. During the Magdelenian period, 17,000 years ago, *Homo sapiens sapiens* decorated the walls of the caves in which he lived. Eight thousand years ago, during the Neolithic period, modern man became sedentary, became a farmer and herder, raised megaliths, dolmen and menhirs. Prehistologia has reconstructed a village of man of the Polished Stone Age, enlivened by scenes of their daily life.

A Brachiosaurus

◄ Lacave Caves: Spider's Legs

Martel

Martel's coat of arms bears three hammers – the favourite weapon of the man who defeated the Saracens at Poitiers in 732 and who gave his name to the town. In 1219, Martel obtained its freedom from the Viscount of Turenne and, times being none too certain, surrounded itself with ramparts, which it punctuated with watchtowers such as the Tournemire tower and heavily fortified gates.
The heart of the town is Place des Consuls where, on the side of the 18th century covered market, the town's old grain measures can still be seen. The Hôtel de la Raymondie, a fine grey stone building, features an elegant series of pointed arcades and is flanked by a thirty-three metre high belfry. Its inner courtyard is decorated with four-lobed rose windows. Built at the end of the 12th century, it was once the home of the Viscounts of Turenne, and today houses the town hall. To the south of the square, the top of a round tower rises above the roof of a house that boasts a succession of sculpted windows with gently rounded contours. This is the Hôtel Fabri, where, in a state of deep despair, Henri Court-Mantel died. Having sided with his mother, Aliénor of Aquitaine, against his furiously angry father, Henri Plantagenet, Henri Court-Mantel laid waste to the Viscounty of Turenne and Quercy and – revolt being an expensive business – shamelessly pillaged the treasures of its abbeys. But the blow he struck against Rocamadour proved to be fatal, and, struck down himself by a terrible fever, he died. Saint Amadour, it seems, had his revenge!
Lined with handsome Renaissance buildings, Rue Droite leads to the sturdily built Saint Maur's Church, whose imposing volume dominates the town. Heavily fortified, its bell tower is a keep, and its long nave ends in a solidly reinforced chevet. Only the cheerful foliage on the sculpted tympanum brings a little relief to its otherwise austere facade.

Perigord Walnuts

Walnut trees have been growing in Perigord and Quercy since ancient times, since walnut shells have been found in certain sites occupied by Cro-Magnon man, seventeen thousand years ago. In the chalky-clay soils of the region, walnut trees, growing twenty to twenty-five metres high, have found favourable ground, as long as the place is not exposed to too much wind or frost. Walnuts were appreciated in the Middle Ages. Their oil was used in food, but also in soap or as fuel for lamps. Walnuts were thrown at newlyweds, their union symbolized by the two sealed parts of the walnut shell. Walnuts are rich in nutrients, with polyunsaturated fatty acids making up 72%, of which a third consists of Omega 3, but also vitamins B1, B2 and E, potassium, zinc, copper, phosphorus, magnesium and potassium. Since the 2002 harvest, Perigord walnuts have been given the AOC quality label, which applies to 7,000 hectares of orchards in the departments of Lot, Dordogne, Corrèze and Charente. There are different varieties: the Franquette, with its thin, pointed shell, is picked in the Lot section of the Dordogne valley; the Marbot, with a fragile shell, is produced in the south of Corrèze. The Corne, whose walnut is hard to extract, grows in Green Perigord, while Black Perigord is the land of the Grandjean, with its perfumed walnuts. The AOC label concerns three products: fresh walnuts, dried walnuts, and green walnuts. Fresh walnuts, harvested in mid-September, are separated from their green husk. Dried, and placed in a dry, dark place, they keep for the entire year. Green walnuts (cerneau) are ready to be eaten. Itineraries forming the Route of Perigord Walnuts introduce travellers along them to walnut growing areas and their products.

Carennac

On the left bank of the Dordogne River, Carennac has retained its old houses with exposed stone walls and dark-tiled roofs, gathered around the former priory and its neighbouring castle. Carennac and its dependencies were ceded to Cluny by the bishop of Cahors in 1040. Odilon, the abbot of Cluny, founded a priory there named for Saint Peter. It was raised to a deanery in 1295 by a papal bull from Boniface VIII. In 1650, François de Salignac de La Mothe Fénelon became the priory's dean. In 1681, he was succeeded by his nephew, François de Pons, the great Fénelon, man of letters, who was elected to the French Academy in 1693, named archbishop of Cambrai in 1695, at the age of forty-four. As a child, he had stayed in Carennac several times. Perhaps it was on the nearby island of Barrade, isolated by the waters of the Dordogne and assimilated by Calypso Island, that the author found the inspiration for his work, *The Adventures of Telemachus, Son of Ulysses* (1699). In 1788, the last four monks were chased from the priory and in 1793, religious services were forbidden in the church until their restoration by the concordat of 1801. Saint Peter's Church opens by a portico topped with a tympanum sculpted in the 12th century. In the centre, within a mandorla framed by symbols of the tetramorph, Christ blesses the faithful. On either side, the apostles, on two superposed registers, converse two by two. On the archivolt surrounding the tympanum, there remain only two dogs and a bear of the sixteen animals that used to decorate it. The Flamboyant Gothic cloister, from the 15th century, flanks the church on its southern side. The southern gallery has kept two Romanesque arcades. The former chapter house, opening onto the eastern gallery, contains a magnificent Entombment of Christ from the 16th century. Joseph of Arimathea and Nicodemus hold the shroud on which Christ is lying, before his tearful mother supported by Mary Salome and Saint John.

Carennac, tympanum of the priory: four apostles (detail)

◄ Carennac, tympanum of the priory: Christ in glory surrounded by symbols of the Evangelists (detail)

Castelnau-Bretenoux

The bastide of Bretenoux was founded by the Lords of Castelnau in the 13th century. Its streets are laid out like a chessboard and the whole bastide is surrounded by ramparts of which only the fortified gates remain today. In Place des Consuls, there is a charming little house with a turret and a columned porch. In return for the protection afforded by the vast fortress below which it sheltered, the little town of Bretenoux remained constant to its Lords. The men of Bretenoux fought at their masters' sides against Simon de Montfort in the Albigensian Crusade, and battled against the English during the Hundred Years' War.

It was from Bretenoux that, as a child, the writer Pierre Loti first gazed upon the "enchanted castle" of Castelnau, "one of the most sumptuous in all feudal France".

And this huge fortress does, indeed, bear an almost disconcerting likeness to the perfect storybook medieval castle born of any child's imagination. At sunset, as the handful of houses in the village of Prudhomat nestle on the gentle slopes below, Castelnau Castle's high red stone walls catch the last rays of sun in a blaze of intense colour.

Castelnau Castle is triangular in form. Joined by curtain walls, two massive round towers and an enormous square keep stand guard at its angles. Running along the top of its battlements, delicate broken lines of machicolations faithfully trace the contours of the castle's blueprint. Solid ramparts support the great terraces, built as a stronghold, and, surrounding the castle, a moat of exceptional width and depth lies hidden behind what remains of the castle's last line of fortifications.

Although much altered over the centuries, Castelnau still retains a very definite air of the imposing medieval fortress it was built to be. The original castle goes back at least to the 11th century, to the time when the Lords of Castelnau were beginning to establish them-

New Towns

During the second half of the 13th century and the first half of the 14th, Aquitaine and Languedoc saw the development of *bastides*. The legal foundation of these new towns was a contract between a lord and one of his vassals. In exchange for protection by the lord, the vassal agreed to pay him part of the revenues drawn from the bastide. In the founding charter were listed the advantages granted to inhabitants coming to settle there. They were exempted from certain taxes and obligations, such as collection, tallage and the right to accomodation. A serf was given his freedom, although serfs of the founding lords were excluded from this measure. Access to property was facilitated. Houses were built on lots marked on an urban planning project, some of whose elements are common to many bastides. In the central square, often surrounded by a covered way, there stood a market hall where markets and fairs took place, as well as a common house where juries or consuls gathered. Representing the inhabitants, they administered the bastide with the *bayle*, the agent of the lord or the king. The plan of the various streets – cart-roads, cross streets, or simply alleys – , often straight lines, were generally in a checkerboard. The bastide's main purpose was economic since, as a place of production and exchange, it settled populations.

They also had a political purpose in this area torn between Plantagenets and Capetians. Nowadays, there remain between three hundred and three hundred and twenty towns or villages that were bastides in the Middle Ages. The main ones of Quercy besides Bretenoux are Rudelle, Montauban, Beauregard, Septfonds, Puylaroque, and Castelnau-Montratier.

selves as the representatives of one of the most powerful noble lines in Quercy.
Then, in 1184, Castelnau's feudal lord, the Count of Toulouse, ceded Castelnau to the neighbouring rival Viscount of Turenne – whose castle was only a good horse ride away – and murderous rivalry broke out between the two clans. Eventually, the feud was settled by the rather fanciful intervention of their common sovereign lord, King Louis VIII, who ordered the Lords of Castelnau to pay homage to their rightful overlord, the Viscount of Turenne. Each year, with great pomp and ceremony, they were to send him the symbolic tribute of their obedience – a single egg! Faithful to the King of France, the Barons of Castelnau fought fiercely against the English in the Hundred Years' War, but Castelnau was not to be spared, and it fell to the English in 1356. Eventually, however, the English troops agreed to abandon their position in return for a large sum of money granted by the ruling assembly – the States General of Languedoc.

The 17th century gallery that runs around the castle's courtyard is all that remains of the buildings of that period, destroyed by fire in 1851. In 1896, the light opera singer, Jean Mouliérat, began restoration of the castle's east wing, fitting the rooms, such as the Hall of the States General of Quercy, and the Salle des Etains, or Pewter Hall, with furniture and other objects dating from the end of the Middle Ages to the 18th century. The oratory contains a 15th century stained-glass window of the crucifixion, from Quimper Cathedral, a polychrome alabaster bas-relief, and a 15th century retable of the Spanish school, illustrating Saint Bartholomew's martyrdom.
In the castle chapel, a strange statue of a gaunt-faced monk stands at the foot of the staircase leading to the Baron's Gallery.
Below the castle, Saint Louis' Collegiate Church, built by the Lords of Castelnau in the 15th century, houses a rich treasure of stalls, stained-glass windows and statues, including one, in polychrome stone, of Christ's baptism.

Castelnau Castle: Chamber of the Quercy States General

2

3

GRAMAT CAUSSE

Saint-Céré

Saint-Céré developed along the shores of the Bave River, a tributary of the Dordogne, at the foot of a sturdy hill, already occupied during the Gallo-Roman period, and crowned with a castle in the 8th century. The Saint-Laurent towers, two large, square keeps, one from the 13th century, the other from the 15th, are surrounded by a wall running along the top of the slope. A more recent main building stands between them and in 1948 it became the refuge of one of the great artists of the 20th century, Jean Lurçat. Several of his tapestries are exhibited at the Casino gallery, east of the town.
Saint-Céré was given a charter of communal liberties in the 13th century by the lords of Turenne, and became wealthy through its contact with three great regions: Quercy, Limousin and Auvergne. In the old town, fine 15th century houses, but also some from the 16th and 17th, attest to this prosperity. On Place du Mercadial, the 15th century house of Jean de Séguirier is flanked by a stone stall, the *taoulié*, on which fishermen sold fish from the Dordogne river. The Ambert mansion, from the 15th century, opens through a Renaissance door onto the rue du Mazel, joined to rue Saint-Cyr through Lagarouste Passage, nearly covered by the corbelled constructions of the houses lining it. In this street, the Miramon mansion, from the 15th century, stands out because of its corner turret. Opposite the chevet of Sainte-Spérie Church, from the 17th and 18th centuries, stands Puymule House, dating from the 15th century. To the east of the town, the pilgrimage chapel of Notre-Dame-de-Verdale nests at the beginning of the gorges of the Tolerme River, a torrent tributary of the Bave, whose flow is controlled upriver by a dam forming a pretty artificial lake.

Losing One's Head

History and legend are often a good mixture, the marvels of the latter sweetening the bitterness of the former. Sérénus, the duke of Aquitaine, had two beautiful children: Spérie, a sweet and charming young girl, whose Christian faith was so strong that she considered giving her life to God, and Clarus, a sturdy boy, proud of his masculine prerogatives. Contemptuous of his sister's vocation, he promised her in marriage to the lord of Castelnau. The power and domination of a lineage often rested on foundations of matrimonial alliances. On the day of her wedding, the girl's only solution was flight. But she did not count on her vexed brother's brutal vengeance. He reacted terribly. She was pursued. Once caught, young Spérie was beheaded on the spot. And in the great tradition of the miracles that are dotted throughout the religious historiography of those distant centuries, the extraordinary posthumous destiny of the young saint was to fire the enthusiasm of generations of faithful ready to believe history as it was recounted at the time. The little saint allegedly picked up her head in her arms and bathed it gently in the clear water of a fountain. Saint Spérie, dead for her love of God, is the patron saint of Saint-Céré. The parish church is named after her. Its chancel contains the tomb of François Maynard, who died in 1646. He was a poet at the court of Louis XIII, and was elected to the Academy by the will of Richelieu. Maynard often spent time in Saint-Céré, enjoying the pomp of courtly life at receptions organized at the neighbouring castle of Castelnau.

The coloured world of Jean Lurçat

Jean Lurçat was born in the Vosges in 1892. The unfortunate risks of the Second World War and the courageous engagement of a man in the ranks of the Resistance were at the origin of Jean Lurçat's attachment to the Lot department. Here he lived through the dark hours of the German occupation – first in Lanzac, then in Souillac. At the end of the war in 1945, he decided to settle at the top of Saint-Laurent hill, which he had admired from afar during his clandestine period. For over twenty years, the high towers were to contain cartons, canvases, and preliminary studies of this genial master tapestry-maker. When Jean Lurçat died in 1966, his wife decided to make the place he had loved into a museum to the happiness that still infuses each of his works. To enter into the museum-workshop of Tours-Saint-Laurent is to discover not only a set of extraordinary works, but it is also to walk side by side with Jean Lurçat and explore this place that he created and the incredible wealth of his creation. It makes us appreciate the brilliant conjunction of talents required to make a successful tapestry. From the cardboard sketch that is already a masterpiece, to the flamboyant delight of the weaving on which the pattern is reborn in full splendour, Lurçat gave tapestry its deserved place in the artistic creation of the 20th century. But even beyond the great tapestries, his talent bursts forth in the brilliant drawings on the beams and the walls. He tried everything and managed everything: ceramics, porcelain, cloth, and furniture all carry his identifiable mark. The symbolism of his tapestries with their pure shapes, the brilliant colours with jagged outlines in deep black, always return to the purity and sparkling harmony of a world of peace and fraternity.

Two Afghan Hounds, painted cardboard

Ceramic plate

Ceramic couscous dish

Roo-star, 1963,
tapestry

The Incongruous, 1965,
tapestry

Montal

Crowning a hill overlooking the Bave valley, Montal Castle is, to all appearances, a medieval fortress, with the wings of its main building set at right angles, and its great round towers tapering into cone-shaped roofs. Appearances, however, can be deceptive, and the only medieval part of the castle is, in fact, its plan, in the form of a feudal citadel, while its decor is quite distinctively Renaissance. The castle was built between 1523 and 1534 by Jeanne de Balzac d'Entraygues, whose elder son had, to her great regret, gone off to the wars that raged in the distant and enchanting lands of Italy. In honour of her cherished child, this woman of sound taste began to build a palace of astonishing elegance. The violent, albeit predictable, death of her son in Italy brought a sudden end to her creativity. Overwhelmed by grief, she had the sad motto "Plus d'Espoir" – "No More Hope" – carved on the pediment of one of the castle's windows. Although the castle was never completed, a splendid house still remains, thanks mainly to the dedicated restoration work undertaken, at his own expense, by Maurice Fenaille at the beginning of the 20th century.
On the edge of the roof, the facade is fringed by a thirty-two metre long frieze where swarms of birds and graceful lovers are tangled together with coats of arms and the initials of Jeanne and her son Robert, trimmed with ribbons of flowering garlands. On the first floor, there is a very pleasing arrangement of alternating mullioned windows and niches with sculptured pediments, each one holding the bust of a different member of the Montal family. Four dormer windows blend the facade harmoniously into the elegant roof.
Inside the castle, a splendid white stone Renaissance staircase leads to the upper floors. Built in an open and intricate style, with an incredible density of delightful sculptures, the staircase is a remarkable and rare work of art.

Montal: castle facade

◄ Montal: the castle

Presque Cave

Beneath a drab and hostile outer shell, the rock can sometimes hide such sparkling jewels that it is hard to imagine how so rough a case can contain so rich a treasure. The cliff that overhangs the Clermont-Ferrand to Cahors road, opened in 1825, seems to be little more than a huge pile of sediment, its origins lost in the complicated twists and turns of long past geological ages. But at the foot of this unpromising cliff, excavations have revealed a remarkable treasure trove – Presque Cave.

Opened to the public in 1922, and cleverly lit, the thousand folds of this fine crystal-lined cavity provide a spectacle of magical splendour. The cave is small in size – no more than 350 metres long – and it branches out in all directions, dividing into a complex series of chambers and galleries.

This cave – where the fluid movement of graceful draperies is stiffened in an instant into solid rock, and where slowly evolving stalagmites leap in tumbling cascades – confuses all normal perception of space, volume and time. Every formation and even the very rock itself seems to gush, soar, leap, bounce, slide and finally melt away in mysterious stillness. Two long columns, nearly nine metres high, elegant and slender, known as the *Candles*, join floor to ceiling in myriad folds of mineral blisters. The bulging form of *Hercules' Columns*, its surface boiling with frozen bubbles, transforms the cave into the imaginary haunt of the great heroes of ancient mythology. Contemplating a magnificent cascade of pearly translucent rock that covers an entire rock face, it is just possible to imagine Antinea, Queen of Atlantis, walking in these enchanted palaces.

Last of all, at the top of a column that spreads in graceful swoops and scrolls, there is a remarkable formation, known as the *Black Madonna*. With her pretty hood of grey rock, the small transparent silhouette of this ethereal Virgin Mary seems to watch and wait.

Presque Cave: Columns of Hercules

◀ Presque Cave: the great column

Remote villages of Upper Quercy

The northern limit of the Gramat causse is the course of the Dordogne River. The causse plateau, with its average height of three hundred metres, sometimes brutally subsides near the river and offers various views: a landscape of abrupt cliffs, rocky cirques or else steep valleys. In this broken-up area, not always easily accessible, villages have developed far from the bigger roads. Examples are Autoire and Loubressac that, half-way between Saint-Céré and Padirac, have retained all their charm. Opposite the cliff containing Presque cave, Autoire has naturally integrated the small river in whose valley it nestles. Upstream from the village, a spectacular waterfall over thirty metres high takes the river across the cliff of the cirque to which it gives its name. The entire site is visible from the belvedere built near the waterfall and from the path leading from it. In the village of Autoire, the church with its squat, square belfry retains a lovely Romanesque chevet. Spread all around it in a very homogeneous whole, stand half-timbered houses with pretty, multi-coloured, extremely steep roofs, some of which are cantilivered, and a small 16th century manor-house with a tower and a bartizan. The sound of running water in a fountain only serves to emphasize the quality of the silence so rare nowadays. Loubressac is a perched village overlooking the valley of the Dordogne. From it the view extends to the castle of Castelnau, the castle of Montal, to Saint-Céré and even to Turenne. Today the village appears to be asleep, yet its remarkable position placed it at the heart of the struggles that tore apart the province over the centuries. In turn pillaged, rescued, delivered, purchased, it managed to survive the darkest hours of the Hundred Years War. Nor was it spared the terrible destruction of the Wars of Religion. The tympanum of its church still bears the scars of those pointless mutilations. A stroll through the quiet little streets, lined with houses of beautiful light-coloured stone, is a charming journey into the scenery of another time. At the foot of the cliff, the castle, which is more a manor-house than a fortress, dates from the 15th century, but was much transformed during the 17th century.

Autoire cirque

Loubressac Castle

Village of Autoire

The Gouffre de Padirac

The gouffre (chasm) of Padirac is an enormous natural well more than 75 metres deep, and 32 metres in diameter, sunk into the deep layers of the limestone Gramat causse. It results from the collapse of the roof of an underground gallery. The huge cone of scree that fills the bottom of the chasm testifies to the size of the collapse that opened this column of light to the sky.

Elevators slowly take the visitor down into the well, then to the foot of the cone of scree. The bright circle of daylight at the surface of the causse slowly recedes. Stairs then lead to the entrance of the gallery of the Source river, cut out of the rock by the underground stream, three hundred metres underground. A walk of three hundred metres leads, along the river, to the pier of the Plane River. The visitor gets into a flat-bottomed, unsinkable boat, led by a guide. Only the splashing of the scull disturbs the chasm's profound silence and the perfect smoothness of the water. The depth of the river varies from forty to fifty metres. The water temperature is stable at 10.5°C; that of the cave remains constantly at 13°C. After nearly seven hundred metres of this fairytale journey, the *Grande Pendeloque* (large pendant), the sharp point of an enormous stalactite hanging down from a height of over seventy-five metres, skims the calm waters of the *lac de la Pluie* (Rainwater Lake) in which it is reflected. From that point on, the journey continues on foot, passes beyond the narrowing of the *Pas du Crocodile* where, on September 28, 1890, during his third exploration, Edouard Alfred Martel got stuck with his canoe, the *Crocodile*. Martel had to wait several hours, hanging on to a stalagmitic pillar, before he was rescued. The narrow path then goes near the *Grand Pilier* (large pillar) with its forty metre height and enters the large chamber where the river widens to form the Grands Gours

The Legend of Saint Martin

Surely, such a great hole, lost in the dark folds of rock, could only be the dreaded lair of the devil himself! For many years, the presence of this huge chasm could be explained only as the work of the Evil One, since, in the eyes of the local population – certainly deeply religious and perhaps even a little superstitious – only he could be capable of opening such a gaping hole down towards hell. And so the legend was born and, nourished by the very strangeness of the site, it grew up to tell of Saint Martin, tirelessly traveling the world in search of lost souls. This, it would seem, was no mean task, since there were any number of lost souls, while happy ones were few and far between. A glance into the sacks of the two rivals would have shown the devil's full to overflowing with the wickedness of the world, while that of the good saint remained depressingly empty. One day, the saint's mule, a courageous beast as a rule, stopped in its tracks, unable or unwilling to go on. Being a good Christian mule, it had, of course, sensed the sulfurous presence of old Lucifer. The devil, all grimaces and jeers, scoffing at the poor saint, proposed a deal – of diabolical simplicity. If the mule could clear the most terrifying obstacle the devil could think of – the yawning chasm he had just opened up at its feet – Saint Martin could have his heavy sack of damned souls. With a prodigious leap, the mule flung itself out over the precipice, clutching at the opposite side with such force that the marks of its hooves can still be seen in the rock today. Thus, Saint Martin, with God's help, saved the souls of the damned. And the devil, shamed and aggrieved, fled muttering down to the depths of hell through the very hole he had just created.

◄ Gouffre de Padirac: waterfall on Upper Lake

Lake, a hundred and twenty metres long and twenty-seven wide. At the end of the lake, the river drops six metres and continues its underground path until it appears at Fontaines de Saint Georges and the Lombard, in Montvalent cirque on the shores of the Dordogne River. From the shore of Grands Gours Lake, a staircase carved into the stone leads into the *Grand Dôme* chamber, where *Lac Supérieur* (Upper Lake) has formed, fed solely by water that has infiltrated from the roof. This roof is ninety-one metres above the surface of the lake, which itself is located twenty metres above the level of the Plane River. The return to the pier goes through a loggia overlooking the *Pas du Crocodile* and provides the best angle from which the visitor can appreciate the impressive size of the *Grand Pilier*.
The Gouffre de Padirac has been open to the public since November 1898, and it was officially inaugurated by two ministers on April 10, 1899. Impressed by what he had explored, and wishing to promote speleology, Edouard Alfred Martel very rapidly conceived the project of developing the chasm for tourism. He acquired it on December 7, 1895. The official deed contains a very curious clause. A persistent legend claimed that the English, before being "thrown out" of French lands at the end of the Hundred Years War, had hidden a treasure in the depths of the chasm. The sellers demanded that they receive a part of this treasure if it were ever discovered. The money needed to develop the chasm still had to be found. Luck turned out to be with him when Martel forgot in a cab his briefcase containing information about this project. The briefcase was returned to him by George Beamish, the heir to a large Irish brewery, who immediately offered to invest in the operation. The financier became the first president of the *Société Anonyme du Puits de Padirac*, which has since been changed into the *Société d'Exploitations Spéléologique de Padirac*.

The Father of Speleology

Edouard Alfred Martel was born in Pontoise in 1859. He was a lawyer but passionately interested in chasms and caves. The Causse area, with its karstic reliefs, was his favoured region. By 1883, he had started exploring sinkholes and caves. In 1888, he explored the galleries of Dargilan cave in the Black Causse. A year later, he was the first to enter the underground network of the Padirac chasm. He led nine expeditions into it. In 1901, he published *Le Gouffre et la Rivière de Padirac*. At that time, tourists had been able to visit the famous chasm without danger for three years. In 1897, with Louis Armand, a locksmith from the village of Rozier, he discovered the sinkhole that he named after his friend and guide. And in 1905, it was Martel who travelled through the Verdon gorges in Haute-Provence through the bottom. His passion for the underground world led him to Slovenia, where he explored the caves of Postojna, to Majorca where he went to Drac cave, but also to the Peloponnese, the Caucasus, England, Ireland and the United States. With less results, he was also interested in decorated caves like that in Font-de-Gaume, in the Perigord. This "speleologist", even before the name was coined in 1890 by the paleontologist Emile Rivière, was the author of many publications that, while extolling the beauty of "unknown France", were strict in their scientific exactness. Martel taught underground geography at the Sorbonne. He died in 1938, but his succession was ensured. In that same year in Padirac, R. de Joly explored four hundred new metres of the underground river, and in 1947, the water of the river was dyed with fluorescein and appeared eleven kilometres away, in Montvalent cirque, near the shores of the Dordogne River.

Gouffre de Padirac: Rainwater Lake seen from the Plane River ▶

Rocamadour

Located in the heart of Gramat causse, the rocky site of Rocamadour rises between Heaven and Earth, in the narrow valley of the Alzou River. This extraordinary site has been occupied by man since the Upper Paleolithic era, as has been proved by the paintings in the Merveilles cave in the causse. In the early days of Christianity, a man of God named Amadour retired to this arid and wild causse, where he built a small oratory dedicated to the Virgin and which was to retain his name. About the year 1000, the chapel of Our Lady of Rocamadour was served by a monk from the abbey of Marcilhac-sur-Célé. It had already become a pilgrimage site, as attested by the presence of *catenulae*, small chains symbolizing their sins, carried by pilgrims around their necks and which were removed at the end of the pilgrimage as a sign of pardon. At the dawn of the 12th century, the pilgrimage grew in size under the influence of Eble, the abbot of the Cluny monastery in Tulle, who, with the help of his brother, the powerful Viscount of Turenne, founded a priory and took over the organization of the pilgrimage. The monks of Tulle occupied the site by 1113-1123, but not without coming into conflict with Marcilhac Abbey, which finally had to give up its claims in 1193. Under the leadership of Géraud d'Escorailles (1152-1188), the successor of Eble, these monks built the great church of Moustier – currently Saint Saviour Basilica, supported by a lower church, today the crypt of Saint Amadour. While an abbot's palace and many chapels were being built, the town developed at the same time, with hostels and shops where pilgrims could buy a *sportelle*, the medal showing the sign of the Virgin, proving that they had accomplished the pilgrimage. During the 12th and 13th centuries, classed at the top of "minor pilgrimages" imposed by ecclesiastical or

The Odour of Sanctity

On May 2, 1244, a simple pilgrim on foot, his pouch slung over his shoulder and his staff in hand, climbed the steep steps of Rocamadour's *via Sancta.* This was in fact King Louis IX, aged thirty. To get there, he had followed the pilgrims' path leading to Santiago de Compostela, passing through Saint-Léonard-de-Noblat, Collonges-la-Rouge and Martel. He was accompanied by his mother, Blanche of Castile, and his three brothers – Charles of Anjou, Alphonse of Poitiers, and Robert of Artois. The local ecclesiastical and temporal authorities were present to celebrate the event. The very independent Viscount of Turenne was at the side of the Bishop of Cahors and of Pierre Vart de Malemort, Abbot of Tulle and of Rocamadour. Blanche of Castile and the future Saint-Louis, both deeply religious, did this pilgrimage for several reasons. Blanche was the daughter of Alphonse VIII, the King of Castile and a benefactor of the Quercy sanctuary, and the granddaughter of Henry II and Aliénor of Aquitaine, who had come on pilgrimage to the shores of the Alzou River. As for her son, his devotion to the Black Virgins, Our Lady of Chartres, Our Lady of the Puy, and of Rocamadour, was well known. Furthermore, his wife Marguerite of Provence, after ten years of marriage, had given him only two daughters. The continuity of the Capetian dynasty had to be ensured. Among her many other miracle-working virtues, Our Lady of Rocamadour possessed that of making women fertile and protecting children. The King of France's prayers on this day of May 1244 must have been heard since less than a year later, on April 30, 1245, a son was born – his successor, the future Phillip III le Hardi. Louis IX was canonized in 1297 by the pope, Boniface VIII.

even civil courts, Rocamadour received many pilgrims guilty of bodily harm, robbery, threats and other offences. Nevertheless, during the entire Middle Ages, other famous saints, princes, bishops or abbots also came to honour the sanctuary with their presence, such as Bernard de Clairvaux in 1147, and the King of England, Henry II Plantagenet, accompanied by his chancellor Thomas à Becket, in 1159. Henry II returned to Rocamadour in 1170, several months before having Becket assassinated in the cathedral of Canterbury, where he had become archbishop. In 1211, Simon de Montfort, in the midst of his crusade against the Albigensian heresy, after receiving in Cahors the homage of several lords of Quercy – Bernard de Cardaillac, Bertrand de Gourdon, and Ratier de Castelnau – met the papal legate in Rocamadour. Dominic, fighting the same battle as Montfort against the Cathar heresy, but spiritually, spent a night praying before the Virgin of Rocamadour. Louis IX and his mother, Blanche of Castile, came as simple pilgrims to the shores of the Alzou in 1244. In 1166, a believer emitted the dying wish to be buried under the parvis of Our Lady's Chapel. When his tomb was dug, a complete body was discovered and identified as being that of Amadour. This discovery increased the fame of the pilgrimage. From then on, pilgrims arrived not only to honour Our Lady, whose miracles were recounted in a hundred and twenty-six accounts in the *Livre des miracles de Notre Dame de Rocamadour* in 1172, but also to venerate the relics of Saint Amadour, the founder of the sanctuary. Amadour was rapidly assimilated to Zacchæus, mentioned by Saint Luke. This short man, a tax collector in Jericho, upon hearing that Jesus was in the town, climbed into a sycamore tree to be sure to see him. Touched by grace, Zacchæus then promised to give half his goods to the poor. Legend also had it that Zacchæus-Amadour became the Virgin's servant and married Saint

Zacchaeus in the sycamore during Christ's entry into Jerusalem

ICI
FUT DECOUVERT
EN 1166
LE CORPS
PARFAITEMENT CONSERVE
DE
St AMADOUR

Veronica. It was with her that he landed at Soulac, on the Médoc Atlantic coast, bringing relics of the Virgin, before retiring alone to the Alzou valley. Whether a goal in itself, or a stage on the great pilgrimage routes, Rocamadour welcomed Compostela pilgrims on the *via Podiensis* coming from Figeac, those who joined it through Brioude and Aurillac, or even those who had left the *via Lemovicensis* at Bénévent-l'Abbaye or Saint-Léonard-de-Noblat. Since Our Lady of Rocamadour was concerned with providing protection on the trail beyond the Pyrenees, did she not, like Saint James Matamore, lend a hand to the soldiers of the Reconquista? The chronicle of the abbey of Trois-Fontaines tells that the Christian soldiers were on the verge of collapsing in 1212, at the battle of Las Navas de Tolosa, when they brandished the banner portraying Our Lady of Rocamadour and set the Muslim enemy to flight. Alphonse VIII (11158-1214) knew it well and counted on its divine

Rocamadour: Stained-glass window in St. Saviour's church

Walking for God

The discovery of the tomb of Saint Amadour in 1166, and the miracles of Our Lady of Rocamadour, listed in 1172 in the "Book of Miracles", attracted a flood of pilgrims to Rocamadour. It was thus able to compete with the great pilgrimage centres of Christianity: Rome, Jerusalem and Santiago de Compostela. The crowd must be imagined, dressed in the long monk's robe and a wide-brimmed hat, carrying a pouch and pilgrim's staff, they slowly climb, some even on their knees, the steps of the Great Staircase of Rocamadour, to finally reach the sanctuaries hanging onto the sides of the cliff. In memory of the pilgrimage, on the little Place des Senhals (signs in Occitan) in front of the entrance to the sanctuaries, the pilgrim bought a sportelle, a bronze medal with its picture of Our Lady of Rocamadour. These sportelles got their name from the Latin *sportella*, the basket or pouch in which the pilgrim placed his purchase. The beginning of a very lucrative business, they were cast in the nearby rue de la Mercerie. The pilgrimage to Rocamadour was also the punishment inflicted on bad Christians, particularly on repentant Cathars, by ecclesiastical authorities who, paradoxically, made this place of worship into the forced hang-out of many wrongdoers guilty of various crimes. They could be found there, under duress but subdued, during the Jubilees, the great collective pardons occasionally accorded by the Church, when Corpus Christi coincided with the feast of Saint John the Baptist.

Sportelle in Rocamadour

protection in 1181, when he donated his lands of Hornillos and Orbanella, near Burgos. After Las Navas de Tolosa, devotion to Our Lady of Rocamadour became a lasting feature in the Iberian peninsula, particularly along the *camino Francés*, where its monks owned important staging posts for pilgrims in Estella, Burgos, Palencia, Astorga and Salamanca. During the Hundred Years War, the well-defended site withstood the English several times, before they took it in 1369 but caused no damage to the religious town. During the Wars of Religion, on September 3 1562, Protestant troops led by Bessonias and Marchastel, paid by the Count of Duras, took Rocamadour and sacked it. The sanctuaries were looted and the body of Saint Amadour was burned. Only the statue of Our Lady and the miraculous bell escaped the soldiers' rage. The architectural site visible today is principally the result of restoration work undertaken during the 19th century under the impetus of Fathers Caillau and Bon-

Rocamadour, Our Lady's Chapel: miraculous bell

Rocamadour: Church parvis

At the Gates of Legend

Notre-Dame-de-Rocamadour is mentioned many times in history, but it is not easy to distinguish verifiable facts from legends created by man's need for the marvellous during the Middle Ages. The year is 770. Roland, Charlemagne's nephew, offers his sword to Our Lady. But after all, unable to give it away, he takes it back, and in exchange gives the weight of his weapon in silver.
At Roncesvalles, he hurls Durandal, the sword, so that it does not fall into the hands of the Saracens. It comes to land in the wall of Saint Michael's chapel. Today, the sword sticks out of a wall in Our Lady's chapel. It should be noted that this wall is below the fresco of the Annunciation in which the Angel Gabriel appears. In the *Chanson de Roland,* it is Gabriel and Saint Michael who "from the peril of the sea" "take away to Paradise the soul of the Count". It should also be noted that Rocamadour and Roncesvalles are on the same path to Santiago de Compostela, the *via Podiensis.*
It is 1212. Our Lady appears to a monk and asks him to go and carry her banner to Las Navas de Tolosa where the battle that will oppose Christian forces and Moors is in preparation. Otherwise, he will die within three days. The monk does not leave and dies. The father abbot takes the banner to the battlefield. Unfurled, it provides victory to the Christian troops...

Battle of Las Navas de Tolosa, Museum of French History, Versailles

The Black Virgin of Our Lady of Rocamadour

Durandal, Roland's sword, in the rock wall above the door of Our Lady's Chapel in Rocamadour

1

2

homme and Monseigneur Bardou, the bishop of Cahors.

Visiting the Site

Rocamadour is the triumph of verticality. The town rises towards the heavens in four steps, illustrating the famous phrase: "The houses on the stream, the churches on the houses, the rocks on the churches, the castle on the rock." The village is located at the foot of the cliff. The Figuier and Salmon Gates remain from its former fortifications. The houses are beautiful with their walls built of light-coloured large stones and their flatstone roofs. Good examples are the 13th century Pomette house on rue de la Mercerie, and the 15th century mansion in which the Town Hall is located, flanked by a round tower topped with a slightly bulging frame. Concurrently a spiritual link and an object of penitence, the Great Staircase, with its two hundred and sixteen steps, rises towards the religious town of Rocamadour – a set of palaces, churches, chapels and crypts – dug into the sides of the cliff and fitting into each other. The abbey palace, former palace of the bishops of Tulle, has perhaps been overly restored. A long pepper-pot tower, crowned by its covered way and its machicolations, marks the very open angle of its two facades, with their openwork of arched windows. This palace currently contains the Museum of Sacred Art.

Our Lady's (Notre-Dame) chapel is the main sanctuary of Rocamadour. The primitive chapel was destroyed in 1479 by the collapse of part of the cliff. Rebuilt shortly afterwards, it was demolished by the Protestants in 1562. Restored in the 17th century, it was damaged once again during the Revolution in 1793. It opens onto the churches' parvis by a fine Renaissance doorway, whose delicacy contrasts with the rough mass of rock rising above Saint Amadour's tomb, protected by its wrought-iron gate. On

Rocamadour: Abbey Palace

the chapel's outer wall, a 15th century fresco, of which only faded images of two skeletons remain, shows the *Dict des trois morts et des trois vifs* (*Tale of three dead and three living*). In the rock face above is set Roland's famous sword, Durandal. The chapel is dark inside. The upper recess of the ciborium, rising above the 19th century, gilded bronze altar, contains the Black Virgin, Our Lady of Rocamadour. This wooden statue, from the 12th century, covered with strips of silver with the patina of centuries, shows the Virgin Mary, in a slightly rigid sitting position, holding the Infant Jesus on her left knee. A faint smile can be seen on the lips of this Black Virgin with the closed eyes. Is the black colour original? Perhaps the wood and metal have darkened over time, or else, have they blackened from the smoke of candles? Nobody knows the answer. Hanging from the chapel's vault, a 9th century bell rang all by itself each time a sailor in danger begged the Virgin for help and promised to make a pilgrimage to Rocamadour if he managed to survive. This phenomenon has been attested several times in the sanctuary's annals: once in the 14th century, twice in the 15th, five times in the 16th, and twice in the 17th. The first miracle recounted in the second part of the Book of Miracles of Our Lady of Rocamadour is that of three monks saved from shipwreck. Monks of Cîteaux, Beaulieu and Châlons had boarded in Venice on their way to Rome and were caught in a storm on the Adriatic Sea. The monks prayed to the Virgin of Rocamadour, each of them promising to bring her a wax boat as a votive offering if they managed to escape with their lives. The sea then calmed down. There is no doubt that the three accomplished what they had promised to do. As did others who benefited from the intervention of Our Lady of Rocamadour. Testimonies can be seen in the votive offerings, model ships and prisoners' chains decorating the chapel, signs of thanksgiving to the Virgin.

A Familiar Companion

During the 14th and 15th centuries, Europe was visited by three of the "horsemen of the Apocalypse" – war, famine and plague. The murderous battles of the Hundred Years War caused many victims. Army requisitions and looting by idle army rabble emptied storehouses and attics of agricultural products whose harvests, from lands impoverished by not being left fallow, were affected by climatic variations.
The plague that landed in Messina in 1347 caused nearly 25 million victims in several months. Whooping cough, dysentery and influenza were prevalent. In 1418, in Paris, 50,000 people died of smallpox, 5,311of them at the Hôtel-Dieu hospital alone. Therefore it was natural that during such difficult centuries, men attempted to present this death that was such a familiar companion. The *Dict des trois morts et des trois vifs* is one of these representations. Its origin was set in the tradition of Saint Macaire, an Egyptian hermit who allegedly met three young princes going hunting on horseback and who showed them three open coffins containing three corpses of kings. The episode is illustrated in a fresco from 1350 at the Campo Santo in Pisa, Italy. It had been set to verse by Baudouin de Condé during the 13th century as a dialogue between the three dead and the three living, the former warning the latter, "You will be what we are, see yourselves in us beforehand. At the hour of death, only good works count." This scene was sculpted in 1408 on the gate of the Innocents Cemetery in Paris on the orders of Duke Jean de Berry. It is treated as a fresco in many churches, as it is in Rocamadour.

Rocamadour, Facade of Our Lady's Chapel: fresco showing *Dict des trois morts et des trois vifs* ▶

Saint Michael's Oratory is located on the third level of Saint Michael's tower-porch, set in the cliff. The tower rises from the parvis, at the top of the Great Staircase. It was probably built by the monks of Saint-Martin-de-Tulle to protect the sanctuary in case the monks of Marcilhac-sur-Célé showed any inclination to take it back. Only very slightly rebuilt, the oratory remains much as it was at its beginnings in the 12th century. Its apse is barrel-vaulted and decorated with a 13th century fresco in which Christ is enthroned in majesty, surrounded by the four evangelists, while Saint Michael prepares to weigh the souls of the damned. The apse bulges on the outside, presenting the shape of a half-bartizan, above the doorway. Above the half-bartizan, on the outer wall of the tower-porch, an even older fresco, dating from the 12th century, fills a shallow recess covered with three arches. Originally, the work was to have covered a large part of the building and perhaps even beyond it to the primitive chapel of Our Lady and the cliff. The voluntary or natural damages that have struck Rocamadour in the course of its history have left behind only a part of the work. Against the blue of the kingdom of Heaven, scenes of the Annunciation and the Visitation are shown. On the left, the angel Gabriel unrolls a phylactery before the Virgin, whose hands are open in a sign of surprise. A dove, symbolizing the Holy Spirit, is resting on the Virgin's halo. On the right, the Virgin holds her cousin Elizabeth in her arms.

Saint Saviour's Church was raised to the rank of basilica in 1913. It is the largest of all the churches and chapels on the parvis. It dates from the 11th, 12th and 13th centuries, but was often remodelled over the course of the years. The chancel has been moved from its former place at the eastern side of the building to the north, facing the entrance, hence making room for two bays of equal size. The left nave is edged by stands built in 1930 and, behind the altar, there is a poly-

St.Michael's Chapel, fresco: Archangel Michael

St. Saviour's Basilica: 16th century wooden Christ

◄ Rocamadour, St. Michael's Chapel: The fresco of Annunciation

Museum of Sacred Art

The history of Rocamadour, devotion to Our Lady and the popularity of the pilgrimage have brought to this place many religious works of art, some of them locally produced, and others from various churches in the Lot department. Since 1968, the Museum of Sacred Art has received and conserved them, although it lends them out to the religious authorities during certain jubilees or processions. To move among these sculptures, gold plate, paintings and stained-glass is to find oneself in the deeply mystical universe of these people who, from the humblest to the wealthiest, created Rocamadour throughout the centuries. Among the exceptional pieces in this museum, dedicated to the French musician Francis Poulenc (1899-1963), the composer of *Litanies à la Vierge Noire* (1936), we should mention the reliquary shrine of Soulomes, in gilded copper enamelled on a wooden core, dating from the 13th century.

Keystone from Saint John the Baptist's Chapel

chrome wooden Christ on the cross from the 16th century, the Saint Saviour.

Built under Saint Saviour's basilica, Saint Amadour's crypt, in pure Romanesque style, was Saint Amadour's burial place until 1562, when the Protestants burned his remains.

Saint Anne and Saint Blaise's chapels surround a lantern of the dead that, like the two funerary plaques embedded on either side of the basilica's doorway, is a reminder that the sanctuary parvis was for a long time a necropolis. The first is a 19th century reconstruction by Abbot Chevalt, while the second dates from the 13th century.

Finally, nothing remains of the 15th century chapel of Saint John the Baptist, other than the frame of the doorway. A keystone carved with a lamb is on display at the Museum of Sacred Art. The chapel, rebuilt by Abbot Chevault between 1868 and 1872, contains the tomb – restored in the 19th century – of Jean de Valon, a knight of the Order of Jerusalem, who died in 1518 and whose family had the original chapel built.

The visitor should raise his eyes to the sky from the churches' parvis. The purpose of the ramparts on the edge of the cliff had been to control this position and, during troubled times, to prevent the enemy from bombarding the town. Within the walls, the castle, built between 1839 and 1841 by Abbot Caillau, is the residence of Rocamadour's chaplains. It is possible to reach the level of the castle from the parvis by a very comfortable rack car, but also by a steep and shady winding path, that climbs the cliff and along which are set stations of the Cross.

To have a fine viewpoint over the site, by day as well as by night, the visitor must go to l'Hospitalet or else go to the other side of the valley and take the little road leading to the chasm of Cabouy and the hamlet of Couzou. And, on the first Sunday in October, against this majestic decor, innumerable hot air balloons dance a silent and colourful ballet.

On the Fringe of the Cliff

A great sanctuary, like that of Rocamadour, attracted thousands of pilgrims in the Middle Ages. The continuous presence of large crowds required the construction, in the 13th century, of places to lodge them outside the walls, such as the Hôpital Saint-Jean. All that remains of this ancient building on the pilgrims' route is a large gate and Saint John's Chapel. The village which developed all around has the classic name of L'Hospitalet. On the way out of the town, on the road to Rocamadour Castle, the Cave of Wonders (grotte des Merveilles) was discovered in 1920. Chosen by men of the Upper Paleolithic era, in the Solutrean period or 20,000 years ago, as a sanctuary cave, they decorated it with drawings and paintings in which we can recognize, or imagine, a horse or a deer or a hyena, witnesses of the dawn of the history of Art. The mischievous but fearful little monkeys

Egyptian Vulture, Eagles' Rock, Rocamadour

Rail Extravaganza, L'Hospitalet

in the Monkey Forest, on the road to Gramat, live on the ten hectares allocated to them on this arid plateau which somewhat resembles their native Africa. The fish shown at L'Hospitalet Aquarium come from local rivers: trout, carp, pike. Still in L'Hospitalet, Robert Mousseau's amazing creation deserves its name of Rail Extravaganza (Féerie du Rail). With care and patience, this expert in model trains has created a fantastic miniature universe, consisting of entirely automatic rail circuits. Near Rocamadour Castle, trained birds of prey fly freely above the cliff and appear to keep jealous watch over the old stones of the ancient town.

A hyena, Cave of Wonders, L'Hospitalet

Barbary macaques, Monkey Forest, L'Hospitalet

Saint John's Chapel, L'Hospitalet

1

2

Gramat

Between Célé and Dordogne, the Gramat causse is a vast tabular plateau, consisting of solid limestone, at an altitude of about 350 metres. The rock appears at the surface in stony groupings, scattered over the thin meadows where pale green grass grows. The strong oak trees, dark juniper, and sweet-smelling honeysuckle, hang on and resist in this environment with its harsh and wild beauty. The internal movement of these rocky surfaces, cracking due to frost and water infiltrations have sculpted and carved the plateau with gorges, chasms and caves.

The town of Gramat, on the shores of the Alzou River, is a busy little commercial centre. Well located in the middle of the causse bearing its name, it has retained just a few of the relics of its past: the Clock Tower (tour de l'Horloge), an ancient fortified gate, and several fine houses of causse stone. Gramat had been raised to communal rank in 1224 but suffered badly from the ravages of the Hundred Years War and the Wars of Religion.

In the Alzou gorges, at a place where the two cliffs form a narrow passage seven metres wide, the Saut mill was built between 1736 and 1739. It is likely that a primitive mill had already been standing on this spot during the 15th or 16th centuries. The mill whose ruins we can see today rises above a ten-metre high waterfall. At the time, its owner, Antoine Raymond de Foulhiac, had to invest the sum of forty thousand *livres*. The river's water, diverted by a race, moved paddle-wheels which turned four pairs of millstones. The wheels were installed in three water rooms, located below the room with the millstones. The miller lived in a lodging built above a millstone room. The wooden floor has disappeared, but the remains of a chimney and mullioned windows are still visible.

Gramat Animal Park

The aims of the founders of this forty-hectare animal park were clear right from the start. Breaking with the more exotic tradition of classical animal parks, their wish was to create an especially protected natural habitat in which local species of animals and plants could live and be seen. No creatures from distant lands are forced to adapt here, as best they can, to a foreign environment. All the plants and animals are native species and, as such, are perfectly at home in the park. Trees and bushes from the plateau, planted in natural and well thought out groups, provide shelter for the most extraordinary crowds of domestic fowl, of unusual and often completely unknown species. And Bison, auroch cattle, ibex and little Prjevalski horses – living reminders of prehistoric ages – revive faded memories of those ancient days, long since obscured by the thick mists of time.

Gramat Animal Park

In the folds of Gramat Causse

Lacapelle-Marival Castle

Gramat Causse is endowed with interesting sites, besides those of Rocamadour and the Gouffre de Padirac. They must be searched for by detouring along little roads, where they may be hidden in the hollow of a small valley. On the road from Saint-Céré to Figeac, Cardaillac fort, at the edge of Gramat causse and Limargue, is the cradle of the Cardaillac family, who held the fief of Lacapelle-Marival, slightly to the north. In 1294, the village obtained its freedom, which was confirmed by a customary charter in the 15th century. The castle in Lacapelle-Marival consists of a square keep from the 13th century, and a 15th century main building, confined by two round towers. The Gothic style church, the market hall, the fine houses and the well-preserved remains of a fortified gate, are reminders of the importance of this small town that was given marquisate status by Louis XIV in 1645. In 1250, Bertrand de Cardaillac, Lord of Lacapelle-Marival, founded the bastide of Rudelle, with its characteristic checkerboard pattern, to the west of his castle. In case of danger, the inhabitants came to shelter in the strong-room of the fortified church, after crossing a labyrinth of narrow stairways, ladders and trapdoors that were intended to

Romanesque church in Livernon

dissuade the attackers. But, besieged in 1595, it had to surrender to the Protestants. Nearby, on the road to Gramat, the village of Thémines has retained a very rustic market hall and the remnants of an 11th century castle. Set up as an urban community by a charter in 1262, it suffered badly during the Hundred Years War. Livernon sits at the top of a hill south-east of Assier. Around the Romanesque church, old houses are grouped in charming disorder. Two kilometres to the north-west of the village, the Pierre Martine dolmen has perfectly resisted the effects of time, with its flagstone weighing over twenty tonnes. Over seven metres long, it sits solidly on its two vertical stones.

Thémines market hall

Assier

Assier was the famous stronghold of the no less famous historical figure, Galiot de Genouillac, Lieutenant General of the Kingdom of France under François I. Assier Castle, built by Galiot in the style of the beautiful châteaus of the Loire Valley, today bears the sad signs of neglect. At the entrance, the portico which used to house a statue of Galiot mounted on horseback, still bears a salamander – the emblem of François I – on its pediment. The other facade, less austere, boasts a fine old colonnaded loggia.
The walls of Assier's Renaissance church, also built by Galiot, bear glorious witness to their founder's greatness. A long frieze, unfurling into the distance, recounts the illustrious story of the man who was behind so many royal victories, as a tangle of canons and standards, arms and battles, surrounds this impressive hero's funeral chapel in a halo of glory.

Assier: Renaissance Church

« J'ayme Fort Une »

This clever play on words, meaning either "I love dearly only one", or "I love fortune", was devised by the Lord of Assier, Ricart de Gourdon, better known as Galiot de Genouillac. Better known, perhaps, but still quite obscure! In fact, the man on whose tomb was engraved, next to his recumbent figure, the motto "After death, a man's good name lives on", is hardly remembered at all. He was once, however, famous indeed. Starting in 1480, as page to Louis XI, he became Equerry to Charles VIII, Seneschal of Armagnac, Captain of the Artillery of Louis XII, Master Equerry of France, Governor of Guyenne, and finally Lieutenant General of the Kingdom. The list of victories that are credited to his name includes the famous Battle of Marignan, where, in 1515 he used a new invention – the canon – for the first time.

Portrait of Galiot

THE BOURIANE REGION

Gourdon

Located between Dordogne and Lot, Bouriane is a rolling, wooded region quite different from the arid causses surrounding it. The soil, mainly sandy or clay, is covered with thickets, moors and woods.

The light-coloured facades of Gourdon's houses climb up the hill on which the town was built. The boulevards follow the nearly perfect circle of its former ramparts. To understand all of Gourdon's charm, the visitor should enter by the fortified Majou gate and stroll along the former main street, la rue du Majou, lined with corbelled houses. The 13th century consulate contains the town hall and opens onto a 17th century covered way. Saint Peter's church is a massive and somewhat heavy monument whose construction began at the beginning of the 16th century. Its nearly windowless walls, the line of machicolations on the facade, and the massive buttresses are all reminders that it was once fortified. Behind the chevet, the former house of the Cavaignac family opens through a classically-styled door, near Saint Peter's church. Cavaignac, a confirmed revolutionary, voted for the death of Louis XVI, then joined Napoleon 1st. His son, General Cavaignac, was a governor of Algeria, and also organized the terrible repression of the 1848 revolution.

The Gothic nave of the Cordeliers church contains a fine, 16th century baptismal font as well as a superb, sculpted holy-water stoup. West of the town, near Léobard, stand the ruins of Nouvelle Abbey, a Cistercian establishment.

Abbaye-Nouvelle near Gourdon

Cougnac Caves

Cougnac Caves are named in the plural, simply because the site boasts not one, but two fine caves, both of them open to the public. The first cave consists of three chambers, opening one after the other along the bed of a fast-flowing subterranean river, their ceilings bored into potholes by the tremendous force of whirlpools. A profusion of slender and sharply pointed stalactites decorates the dark rock face with a light translucent laciness.

Two hundred metres away, the second cave, discovered in 1949, consists of two main chambers. The strange confusion of structure and decoration in the *Hall of Columns* could almost have been taken from the muddled pages of the sketch book of a slightly potty architect. A single column, formed by the slow meeting of stalagmites and stalactites, seems to bear the weight of the entire ceiling, lending the chamber a certain architectural elegance. The walls of the chamber of prehistoric paintings are covered with hundreds of drawings, some of which were made nearly 25,000 years ago, such as those on the Megaceros panel, and 14,000 years ago for the finger paintings. Over the lovely red-ochre outline of a mammoth standing firmly on stiff, heavy legs, a strange jet-black human silhouette, riddled with lines and arrows, has been drawn. The meaning of this, and other drawings, remains a mystery. Even the shape of the man, with a long body and what seem to be the hindquarters of an animal, is puzzling. Equally disconcerting, is the incredible skill shown by the artists who have worked with and around rock slides, roughness, and natural bumps in the rock face, blending these faults into the drawings themselves, and producing perfect images of, for example, an ibex with long curved horns. There are paintings too of that great deer-like animal, the now extinct megaceros, whose powerful antlers could measure up to three metres across. Staring out across thousands of years, there is a look of great astonishment in their wide eyes.

Cougnac caves: large chamber in the first cave

The Refuge of Ossip Zadkine

The sculptor Ossip Zadkine was born in Russia in 1890. He came to live in Paris where he learned about Cubism. Upon discovering the village of Arques in 1934, he settled there and exhibited his works in the beautiful Romanesque Saint-Laurent church. Zadkine's large, wooden *Christ* expresses all the world's suffering. In the crypt, his Pietà no longer manages to hold her son, whose head hangs in emptiness. Three other works by the sculptor are shown outside the building, including a pathetic *Screaming Harlequin*. In the village, three rooms of the Zadkine Museum present an interesting range of this genial artist's talent. He did not confine himself to sculpture, but also dealt with watercolours, lithographs, tapestries, wood and bronze. In 1954, Zadkine discovered a set of frescoes in the modest church in the neighbouring village of Saint-André-des-Arques. Around the fine window that lights up the chancel, the theme of the Annunciation presents the Virgin and the Archangel Gabriel. On either side, the apostles present their traditional symbols to the faithful. Peter holds the key to Paradise in his hand. James the Greater leans on his pilgrim's staff. Andrew carries the cross of his torture. On the semi-dome vault, decorated with a multitude of stars, Christ, whose image has been somewhat altered, sits on a rainbow with well-rounded arches. He holds the terrestrial globe topped with a cross, the ecumenical symbol of the religion whose prophet he was. In his right hand, he blesses the faithful who have come to pray at his feet. Finally, on one of the pillars of the apse, the Infant Jesus waits for Saint Christopher, on the other pillar, to carry him across the river.

Saint-Laurent Church, Les Arques

Pietà by Zadkine, crypt of Saint-Laurent Church, Les Arques

Fresco of apostles,
Saint-André-des-Arques
Church

The Virgin, fresco of the Annunciation (detail), Saint-André-des-Arques Church

Screaming Harlequin by Zadkine, 1936, Saint-Laurent Church, Les Arques

5

THE LOT VALLEY

Bonaguil Castle

Huge, colossal and powerful. These are just a few of the adjectives that could be used to describe the mighty fortress of Bonaguil Castle, standing guard on the borders of the Agenais, Quercy and Perigord regions. Safe behind its labyrinth of high interlocking curtain walls, and bristling with fortifications, this magnificent castle was hideout and refuge, a proud, impregnable domain whose noble lords reveled in their castle's glorious strength.

However, Bonaguil was not always the perfect fairy-tale castle, drawn as if by magic from some fantastic legend. At a time when people lived in constant fear of a thousand daily dangers, a rocky peak was a godsend. But a needle-shaped rocky peak whose fragile heart had been eroded into a cave, as it had at Bonaguil, was an ideal place of refuge. Bonaguil's name, from "bonne aiguille", simply means "good needle". Following the tortuous line of the rock face, Bonaguil's first fortified walls were raised around a bold pentagonal keep.

Historical records reveal little of how the famous Roquefeuil family came to own this eagle's nest of a stronghold. They do show, however, that by the end of the 14th century, the Roquefeuils were one of a number of noble families who, with a mixture of courage and opportunism, had carved out great domains for themselves in the Midi region of France. Amid the complex foliage of their family tree, one branch bore Jean de Roquefeuil. It was he, and more especially his quick-tempered, hard-hearted son, Béranger, who were to build Bonaguil into a glorious fortress.

Full of charm and intrigue, this strange castle looks in every detail just like the most perfect of all storybook medieval fortresses. In spite of the damage inflicted during the French Revolution – when the pursuit of equality was often pushed beyond the bounds of reason, and attempts were made to destroy all the symbols of the old order – all the typical architectural features that might be expected of such a defensive fortress can still be seen at Bonaguil. Borne forward like an enormous shield, the barbican's protective shell of rock and high rounded walls braces itself defensively in front of the first of Bonaguil's drawbridges. Not content with a modest single drawbridge, this stone monster boasts a total of seven of them, scattered at various strategic points around the whole castle. If ever attacked, Bonaguil would wall itself in behind its last line of defence, to the dull metallic echo of its running drawbridge chains.

Inside the castle, the extraordinary boldness of the architectural design is immediately noticeable. The eye is drawn by the gleaming white of the keep's high prow, to admire the wonderful purity of its outline. The narrow twist of the tower's giddy spiral staircase opens out suddenly onto the keep's upper terrace – a vantage point from which the view of Bonaguil is well worth seeing. From here, high walls of soft grey, faded by wind and rain, twist around the interlacing emerald green ribbons of tender lawn that carpet the moats. This gentle blend of pale stone and vivid grass creates an idyllic setting, blotting out all the memories of suffering, tragedy, danger and death that haunt the castle's terrible past.

The sheer size and ingenuity of the castle's defenses, which made any large-scale works for channeling water superfluous, made Bonaguil one of the best defended fortresses in the medieval western world. Under the castle, its cave, opening like the galleries of a fabulous cavern in the very heart of the rock, was used

both as storeroom and a protected passageway between courtyard and moat.
The narrow passage that opens up inside the needle-shaped peak of rock where Bonaguil Castle is built, was first used, long ago, as a hiding place and refuge. In the Middle Ages, this natural feature was enlarged – using cutting tools that have, to this day, left their mark – into a safe and spacious gallery that forms the heart of the castle's defenses.
Of Bonaguil's magnificent towers, the largest, a circular stony giant, aptly named the Grosse Tour, or Big Tower, measures fourteen metres across, and is trimmed with a geometric border of inverted pyramid-shaped corbels that support machicolations, cut clean and round. Its raw stone walls conceal a swarm of passageways, spiral staircases, secret doors, blind corridors and fine vaulted rooms. Blushing like a pretty maiden, the Red Tower reinforces the sharp angle of the Lord's living quarters. Set deep into the complicated folds of the curtain walls that join these two towers, the Square Tower, protected by a drawbridge and an arrangement of canon loopholes that were cleverly designed to allow canon fire crossways and to the rear, was an essential defensive post. Its heavy door was fitted with the simple but ingenious device of a reversed central hinge, protecting it from any danger of being forced by a well manipulated lever.
Of all the castle's defenses, the *chicane* is probably the most fiendishly cunning. Behind a mild and friendly exterior, it conceals the most terrible trap. A narrow winding passageway, punctuated with doors designed to make progress difficult, led the poor unsuspecting enemy right to the foot of the ramparts, and to the mercy of the canons.
Bonaguil's defenses rest heavily on the use of artillery. Muskets, culverin long canons, harquebus matchlock guns, and batteries of canon were all used to turn Bonaguil into an impregnable stronghold. Yet in spite of its military might and menacing air, Bonaguil is still the legendary castle of childhood dreams.

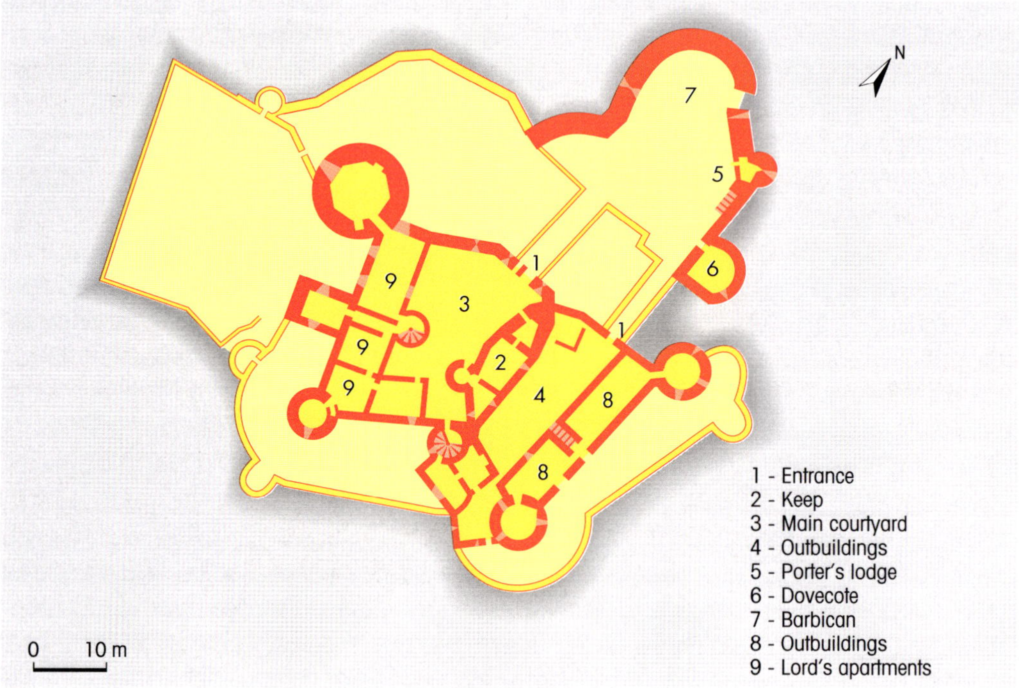

Plan of Bonaguil

The Lot River, from meander to meander

From Cahors to Fumel, the calmed Lot River winds through a set of meanders. Towns and villages mingle in the loops of the river. The presence of humans has been attested in this valley for a very long time with, for example, the Gauls' *oppidum* on the site of l'Impernal. The location, on a height above the curve of a tight meander, was well chosen. At its narrowest point, the isthmus is barely a hundred metres wide. On a spit of land encircled by the same meander, where the headland of La Pistoule rises to the south by the river, Luzech has developed around a 13th century keep, the remains of a 12th century castle, of which Richard the Lionheart became the master. The seat of one of Quercy's four baronies, the town resisted English attacks during the Hundred Years War. The Chapel of the Blue Penitents dates from the 12th century, as does the Consuls House, with its two paired bays. Today it contains the tourist office and the museum presenting the results of excavations of the Impernal *oppidum*. These were begun in 1873 by Castagné and continued by Viré. Downstream, the town of Puy-l'Evêque, like Luzech, took up a position at the heart of a narrow meander, at the point of which, on the concave side, stands Grézels with La Coste castle, built in the 12th century by the bishops (évêque) of Cahors. For centuries, Puy-l'Evêque was a stronghold of the bishops of Cahors. Its place names are a reminder of this period. Only the 13th century keep remains of the castle. A target of the English during the Hundred Years War, the town owes them a defensive system that was much appreciated during the Wars of Religion.

A meander of the Lot seen from Impernal *oppidum*

Luzech and a loop of the Lot River

Puy-l'Évêque

Mercuès

Mercuès, today at the heart of the Cahors vineyards, was, starting in 1212, the residence of the bishops of Cahors. Situated a hundred metres above the Lot River, they had an unrestricted view over the valley and the seat of their bishopric. The current castle was built and fortified in the 14th and 15th centuries, those of the Hundred Years War, during which the French and English fought over it, as did Catholics and Huguenots during the Wars of Religion in the 16th century. Burnt down by the latter, it was restored in the 17th century by a bishop of Cahors. Pierre Habert had taken in eight hundred of his flock to protect them from the plague epidemic that was raging at the time. He used the grateful and available help to carry out the needed repairs and to surround the castle with terraces and gardens. A crenelated keep, pepper-pot roofs, a suspended bartizan, the delicate lacework of the machicolations – all was raised, reinforced and restored. The warlike citadel was transformed into a Renaissance castle, in a refined framework of terraces and cool gardens spread over the gentle slopes. And we can be sure that, even though the antique temple dedicated to Mercury has disappeared, the wing-footed god still secretly finds pleasant surroundings in this Mercuès Castle that is today one of the jewels of the Relais et Châteaux hotel chain. The viaduct with thirty-three arches carrying the Cahors-Brive railway line rests for a kilometre along the base of the cliff on which Mercuès Castle stands. No less than eight years were needed to build this line in the 19th century, even though the distance separating the two cities is only a hundred and eighty-eight kilometres. But there was the causse in between. Eleven railway bridges were built and nineteen tunnels were dug, the first of which, with its two hundred and ninety-four metres in length, crosses the buttress of Mercuès Castle.

A Long Tradition

Grapes were planted in Quercy by the Romans, perhaps after taking the *oppidum* of Uxellodunum by Julius Caesar in 51BCE and marking the end of the Gallic Wars. Jacques Duèse, a native of Cahors and pope in Avignon from 1316 to 1334 under the name John XXII, appreciated the wine of Cahors at his own table as well as during religious services. François I, a man of good taste, had the wine brought to his court in Fontainebleau. During the 19th century, the vineyards of Cahors covered forty thousand hectares. Alas! Phylloxera almost completely destroyed them. Production of Cahors wine was started again in 1947. Hard work, vine by vine, was needed to recover the ground lost. In 1951, Cahors wine was given the classification VDQS (Vin délimité de qualité supérieure). The harsh winter of 1956 nearly brought all this work to nothing but finally, in 1971, Cahors wine earned the label AOC (Appellation d'Origine Contrôlée). This quality label limits planting density to four thousand vines per hectare and wine production to forty-five to sixty hectolitres per hectare. The extent of this Appellation area is from east of Cahors to west of Puy-l'Évêque, on either side of the Lot River. It covers twenty-one thousand, seven hundred hectares, of which four thousand, three hundred are vineyards. The vines grow on the alluvial terraces of the Lot and extend, in the north, to the edge of the causse and on the causse itself. Protected from the cold by the limestone base of the plateau, the area benefits from the climatic regulation of the Lot, which flows between the vineyards. Auxerrois grapes (also called cot noir or malbec) make up at least seventy percent of the composition of Cahors wine, a red, tannic, perfumed, full-bodied and colourful wine, which can be aged even though it can also be drunk young. In the year 2000, its production was over two hundred and fifty thousand hectolitres.

◄ Mercuès Castle

Cahors

It has been such a long time since Léon Gambetta's (1838-1882) native city chose its river, and the river its city, that it seems pointless to mention that the Lot River flows through Cahors, or that Cahors is located on the shores of the Lot! The river winds around the town and each finds in the other the balance that has made this one of the loveliest sites in France.

A resurgence, the fountain of Chartreux, with its clear and abundant waters, rapidly attracted man's attention. The Gallic tribe of the Cadurques created a fine town around it – the old *Divona Cadurcorum*, which became Cadurca and then Cahors. After the Cadurcian *oppidum* of Uxellodunum was taken by Julius Caesar in 51 BCE, the town became Gallo-Roman and urbanized. A forum, thermal baths, temples and a theatre were built. And even though the water from the cool spring was still worshipped as a tutelary divinity, it was no longer the only interest of the valley. Nearby, in the Lot valley, particularly favourable climate and soil fostered the growing of grapevines, dear to Bacchus. During the Middle Ages, Cahors was promoted to being the seat of a flourishing bishopric. The town was then freed from the overwhelming overlordship of the counts of Toulouse. In fact, in June 1211, the town's bishop and lord, Guillaume de Cardaillac, went to Toulouse, besieged by Simon de Montfort, to pay homage to the latter. At that time, Cahors took over the space that the river had always promised it – at a point on one of the bends. A navigable river, a quality product to export and, especially, solid business sense and even usury, gave the city its rank among the greatest towns of the medieval period. During the 13th century, Cahors had nearly 40,000 inhabitants. Traders and bankers from Lombardy came to settle there. They dared to affront the religious in-

The Prince of Poets

Clément Marot was born in 1496 in Cahors, in the Badernes neighbourhood. His father, Jean Marot, was already well-known as a poet. However his son, who knew how to pick his patrons, such as François I whose valet he became in 1526, and Marguerite de Valois, was soon to overshadow his father. But Clément Marot was too frank and unwilling to accept the often obscure dogma of the Catholic Church. He was suspected of heresy in 1527, but his relations at court protected him. He even dared to eat bacon during Lent, and for this extremely serious sin was imprisoned at the Châtelet in Paris. He evoked this misadventure in acid terms in *Enfer* (Hell).

In 1534, he was compromised in the Placard affair – these were posters printed by Protestants, criticizing the "horrible abuses of the royal Mass". He was forced to seek exile outside France and found refuge with Marguerite of Navarre in Nérac, then with Renée of France in Ferrara. He was eventually reprieved. Clément Marot was also a carefree fellow who, in 1535, published one after the other his famous *Blason du beau tétin* and *Contre blason du laid tétin*, a collection of libertine poems that were immediately successful. In a very pure style, this great rhetorician, a translator of Virgil and publisher of François Villon, provided French literature with its first sonnets. The poet also knew how to speak of his homeland, describing his childhood in Quercy in *Épilogue au roi sous le nom de Pan et de Robin*. "In the place that I am speaking of/the Lot River flows its unclear water/that many rocks cross and surround/ to go and join the straight line.of Garonne./ To speak concisely, it is Cahors in Quercy."

The publication of *Trente Psaumes* (Thirty Psalms) in 1541 lost him royal favour. Once again he set out into exile, stayed in Geneva, Chambéry, then Turin where he died in 1544.

◄ Cahors: Chartreux Fountain

terdiction on lending money with interest. Locals soon imitated them. They all became so wealthy that the name *cahorsin* became synonymous with financier and Cahors became the prime banking centre of Europe. In 1270, the city received a communal charter and an underlying rivalry arose between the consuls, eminent citizens chosen to run the city, and the episcopal powers. Jacques Duèse came from a rich family in Cahors. When he became pope in Avignon under the name John XXII in 1332, he founded in his native city the greatest university of the kingdom's south-western area. The young Fénélon was to come to study there later. Half the population of Cahors succumbed to the plague in 1347. During the Hundred Years War, the town remained loyal to the French cause but the Treaty of Brétigny in 1360 handed it over to the English, despite protests from the consuls and the inhabitants. The English abandoned it in ruins nearly a century later. The city's fortifications date from these troubled times. To the north-east, facing Cabessut Island, the barbican is a remnant of ramparts that barred the narrow isthmus cut out by the river. Nearby, Saint John's Tower, or tour des Pendus (hanged) overlooks the Lot. Other towers, symbols of worry or success, raise their high outlines in the old city: John XXII's tower, remaining from the palace of Pierre Duèse, the pope's brother, the tower of the King's Castle, former residence of the Quercy seneschal, both from the 14th century, the hexagonal tower of Pélegry college, founded in 1368, from the 15th century. During the Wars of Religion, Cahors was Catholic and in 1560 its Protestants were massacred in rue des Soubirous. In 1580, Henry of Navarre took over the town and sacked it. But when religious peace returned, Cahors' commercial tradition reasserted itself and great merchants displayed their wealth and their refined taste on the facades of their houses. At number 117 of rue Lastié, a 16th century house opens with pow-

Cahors: Barbican

Cahors: house, rue Lastié ▶

117

erful arcades topped with the light lacework of its three windows with twin bays. A half-timbered, corbelled facade has been well restored at number 156. At number 116, rue Nationale, the door of Bon Pasteur House, a former 17th century hospital, is covered with sculpted scrollwork, opening by a wooden leaf on which fruit, flowers and luxurious foliage spread exuberantly. The Roaldès House is a large half-timbered mansion, topped with a wooden gallery opening under the roof, called a *soleilho*. Built in the 15th century, this is the house where Henry IV is said to have stayed in 1580, during the siege of Cahors. During the 17th century, it belonged to a wealthy, eminent Quercy family – the de Roaldès.

The presence of a primitive cathedral named for Saint Stephen is attested in Cahors at the beginning of the 7th century in the *Vita* of the bishop of Auxerre, Saint Didier. Work on the cathedral was undertaken in the second half of the 12th century and continued in successive steps over several centuries. In 1119, Pope Calixtus II, who had just participated in the Council of Toulouse, consecrated the main altar. The Ascension Doorway was made in 1135. It was moved to the building's north facade under the bishopric (1316-1324) of Guillaume Labroue, the cousin of John XXII, when the latter had the facade rearranged and divided by two towers, giving it the severe appearance of a fortress, somewhat softened by the presence of a rose-window. The bishop Antoine de Luzech (1495-1509) had the cloister built on the southern side of the cathedral. In 1580, Henry of Navarre's Huguenots sacked the cathedral. Two years later, the future Henry IV donated three thousand *livres* to the chapter to pay for reparations. On the northern doorway's tympanum, medieval decorators have shown the Ascension. In the centre in a mandorla, Christ ascends, holding a book in his left hand. He is surrounded by two angels. At the same level, between each angel and the archi-

Cahors: Roaldès House

volt, there are scenes from the life of Saint Stephen. Below Christ, under a trilobate arcade, the Virgin raises her right hand towards her son. On his right and left, under other trilobate arcades, the apostles converse two by two.
Inside, the nave is topped by two domes on pendentives. The western cupola is decorated in its centre with a fresco showing the lapidation of Saint Stephen, surrounded by his torturers, while at the edges is a succession of portraits of eight prophets. The walls of the chancel and the apse are decorated with other paintings. The first shows the martyrdom of Saint Stephen and Saint Spérie, and the second, the coronation of the Virgin and the Adoration of the Three Wise Men. Although the general architecture of the Gothic cloister has not been altered since its construction, the same cannot be said for the sculptures which suffered greatly during the Huguenot raid in 1580, and of which only a few remnants are still in place, such as a pendentive keystone. The small Saint-Gausbert chapel was formerly the capitular room. Currently it is the Museum of Sacred Art and presents liturgical vestments and objects from the Cahors Bishopric since its origins. Furthermore, it contains its own frescoes and paintings dating respectively from the 15th and 16th centuries.
But the symbol of Cahors is Valentré Bridge. It spans the Lot River downstream from the city and used to be part of the fortifications that defended the meander of the river. Well preserved, it has come down to us as one of the most perfect examples of military architecture of the Middle Ages. Construction of the bridge began in 1308 and its role as a fortress was never denied. Even adversaries as determined as the English troops during the Hundred Years War or Henry of Navarre ardently defending his reformed religion were never able to get the better of it. It must be said that the concept of the structure was extremely successful, both from a functional

Cahors, St. Stephen's Cathedral: a gallery in the cloister

Cahors: Valentré Bridge ▶

and an architectural point of view. Its three square, powerful towers are solidly set on stone bases. Nearly blind or simply pierced with thin openings, more than forty meters above the water of the river, they are topped with strong rooms provided, naturally, with crenelations, merlons and machicolations. The two outer towers had gates and a porticullis. The bridge's piers are impressive, anchored on the river's deep bed and protected behind the pointed shield by a crenelated cutwater. The pointed arches that support the face of the bridge are extremely elegant and are reflected in the water with perfect symmetry. Resting on opposite shores, two barbicans protected access to the bridge. Finally, whether legend or truth, the solid cage of infamy exhibited in the Devil's tower is a reminder of the brutality with which adulterous women were treated. Sheltering behind this giant, the little city could peacefully carry on with its active and flourishing commercial activities. Since the restoration by Viollet-le-Duc in 1874, who somewhat altered the original work, a devil carved at the summit of the central tower, on the last stone that it is trying to tear away, evokes the legend of the bridge's construction.

North of Cahors, near Saint-Pierre-Lafeuille, surging out of the deep mass of vegetation besieging it, Roussillon Castle is a majestic ruin. Roussillon's walls bear the traces of the violent attacks that it underwent. Built during the 12th century, it was a part of the series of fortresses and strongholds set all around Cahors, on the domains of the bishops of Cahors whose prime objective was defending the rich episcopal city. Attacked many times, it was taken and retaken. When it became the possession of the famous Perigord family Gontaut-Biron, Roussillon enjoyed a respite, but the French Revolution and the anger of common people who wanted to overthrow the symbols of the Ancien Régime for a time turned the proud castle into a stone quarry. But it resisted all attacks.

Damn it!

Building a colossal edifice like Valentré Bridge requires equally colossal financial and technical efforts. Desperate to finish his bridge, the architect began by calling on the saints of paradise to help him in his task. But when his prayers seemed to fall on deaf heavenly ears, this shrewd and artful man was struck by another idea, certainly more dangerous, but perhaps more effective: if the master of heaven either could not or would not help, then he would call upon the master of hell instead. Satan, ever eager to attend to a troubled soul seeking damnation, rushed to answer his call. Help him? Of course. But at a price. Once the work was completed, the devil, who was very fond of souls, was to receive his reward: the architect's soul. The architect wanted his bridge; the devil wanted the architect's soul, so the matter was settled and work was begun. Arches, piers, parapets and towers shot up with what can only be called devilish speed! The architect was delighted; the devil was laughing! But then the architect began to wonder if any bridge, even one such as his Valentré, was really worth eternal damnation. Deciding that he had no desire to spend the rest of eternity roasting over the fires of hell, he had yet another, rather diabolical, idea. "Devil, fetch me water for my mortar!" he commanded, indicating the recipient – a sieve, riddled with a thousand holes. Although he worked like a demon, the poor devil was unable to finish the bridge and keep his promise, and had to admit he was beaten. Beaten? But if that was the case, why is it that every time the last stone is set firmly in place at the top of the great tower, does it fall away? It is solidly sealed and each time it falls heavily to the ground. Damn it! An ill wind perhaps?

Henri Martin, an Occitan Painter

Henri Martin was born in Toulouse in 1860. He studied at the École des Beaux-Arts in Toulouse, then in that of Paris, where he was the disciple of Jean-Paul Laurens, also from Haute-Garonne. In Italy, Martin discovered the art of Giotto and met Segantini, a Divisionist from Lombardy known for his Grisons landscapes, whose technique influenced him. Martin was influenced by Seurat's Pointilism and by Impressionism. He painted with irregular and parallel touches of matter. He produced allegorical scenes, celestial apparitions, and poetic visions. With *Les Faucheurs* (the Reapers) (1903) he described the world of peasants. This wall painting decorates the walls of the Pas Perdus Chamber at the Capitole (City Hall) of Toulouse, as does the 1906 work, *Les Promeneurs*. Among the strollers on this painting, one can

Young woman, study for *In the Light* by Henri Martin, Henri Martin Museum, Cahors

La Bastide-du-Vert, morning by Henri Martin, Henri Martin Museum, Cahors

Boats in Collioure by Henri Martin, Henri Martin Museum, Cahors

Caricature of Gambetta by Pépin, Henri Martin Museum, Cahors

recognize the figure of Jean Jaurès. Martin received official orders for his paintings. Hence in 1914, he decorated the walls of the law court (Palais de Justice) in Paris with *Travail* (Work), on the world of workers. In 1900, he purchased the Marquayrol domain in Labastide-du-Vert, north of Luzech. During his stays in the Lot, he painted landscapes of his adopted village, where he died in 1943. He donated seventeen of his canvases, including the triptych made for the war memorial of Cahors, to that city's museum, which took the name of the donor. The museum is located in the former bishop's palace, from the 17th century. It also contains a collection of prehistoric and Gallo-Roman pieces. Since 1996, the museum has undertaken the renovation of its works from the 19th and 20th centuries. It contains the Gambetta collection, presenting a selection from the more than two thousand works, and is considering the opening of a Clément Marot section.

Bouziès

The waters of the Lot and Célé Rivers join at Conduché, forming a narrow strip of land. Downstream from the confluence, on the left bank of the Lot River, the village of Bouziès is linked by a bridge to the other shore on which a cliff rises. This cliff was dug out and fortified to be transformed into a troglodytic defence post, known as the English castle (château des Anglais). In 1877, when river navigation was still an economically important activity, the base of the cliff on the left bank of the Lot, between Bouziès and Saint-Cirq-Lapopie, was carved to create the Ganil towpath.
In 1984, the sculptor Daniel Monnier decided to work on the vertical wall of this towpath, which is no longer used. Over thirty-six square metres, a bas-relief fresco shows fish and birds, a reflection of the fauna that enlivens and populates the Lot and its shores.

Bouziès: The English Castle

Bouziès: the cliffs

Wealth Comes from the River

A river is also a major route for trade between upstream and downstream and the Lot River is no exception to the rule. But navigation is difficult here. The Lot has gravel banks and their usual companions, deep whirlpools, high water at times of flood and trickles during endless low water. How then can it be trusted? But the Lot – pronounced with its final t, the phonetic inheritance of its former name of Olt – has been tamed by the inhabitants of its valley. During the 13th century, the bishops of Cahors began to develop the river that meanders through their lands in the direction of the great markets of Bordeaux. The English, when they had become lords of Guyenne in the 14th century, carried on the work. Wine from Cahors, chestnuts, wood, the famous staves for making casks, cheese from Auvergne – the transportation of all these products was difficult over rutted paths. The river was safer after all.

At the confluence of the Lot and Célé Rivers

In the 17th century, Colbert had locks and towpaths built. The coal mines of Decazeville, worked starting in the 19th century, were to provide the river with its golden age. Two hundred and fifty thousand tonnes of coal were loaded annually at Bouquiès and taken to the Garonne River. On the way back, stockfish, *estofi*, or dried cod, was attached to the poops of barges called *gabares*, flat-bottomed boats that had to be towed from the towpaths, and slowly softened in the waters of the river. At its arrival in Livinhac, near

Saint-Cirq-Lapopie overlooking the Lot

Decazeville, it was ready to be eaten in the *estofinado*, a typical dish of the coal country – a subtle mixture of slightly mashed potatoes and small pieces of cod that would melt in the mouth. Roads and railways have definitely put an end to this means of transportation but since 1990, between Luzech and Saint-Cirq-Lapopie, the Lot has once again started transporting people – this time tourists. Seventy-four kilometres of waterways, including fourteen locks, have been opened and offer various possibilities, from excursions in passenger boats to the rental of houseboats.

Saint-Cirq-Lapopie

"Like some impossible kind of midnight rose". This is how the French writer André Breton described his first enchanting glimpse of Saint-Cirq-Lapopie, in 1951. The spell was never to be broken. Here, he said, he could "stop wishing himself elsewhere". Saint-Cirq-Lapopie's great charm is, indeed, undeniable. Once mastered, the name of this village trips off the tongue lightly, like a poppy's falling petals, and it is simply impossible not to fall in love with the place. Set delicately like a precious gemstone between rock and sky, above the soft foamy white murmurings of the river, there are just not enough superlatives in the English language to adequately convey the sweet tranquillity of this marvelous place. Built on the uppermost ledges of a rocky cliff, the village, with its church, like a rose in full bloom, outlined boldly against the sky, looks down onto the Lot, some eighty metres below. At day break, when the soft mists of morning rise from the river to cling in light folds to the cliff, the village seems to float on ethereal waves. In the evening, when every stone bursts with colour in the golden sunlight, the village glows as if with an inner life and, for a fleeting moment, its soul is one with eternity. Saint-Cirq-Lapopie's wonderful and enduring charm lies in this fragile balance between a site of exceptional natural loveliness and man's extraordinary capacity to create beauty. The inhabitants of this place have always been aware of its magical quality. Much too good a defensive position to escape notice, it was already occupied in Gallo-Roman times, as the pottery and mosaics found in the Gallo-Roman villa at Pompéjac recall.

Divided into three parts, its name recalls the two major events that marked the birth of the present village. Saint Cirq, or Saint Cyr, was a young martyr killed, a long way from the Lot, in Asia Minor. Traveling along mysterious paths, his relics finally arrived here, in deepest

Saint-Cirq-Lapopie: 14th century half-timbered house

Quercy. Lapopie was the family name of the local Lords. Their era, marked by sieges and obscure battles, lasted until the 13th century, when the little town's fortunes began to improve. Large houses where built, then a hospital and a fair ground, a church and, of course, the castles belonging to the succession of Lords that governed here. The list of their noble names is long and impressive, and after the Lapopies, includes the Gourdons, the Cardaillacs, and the Hébrards of Saint-Sulpice.

In 1236, the town was granted a Customs Charter, and six governing magistrates, or *consuls,* were appointed. Taken time after time during the Hundred Years' War, the town was finally betrayed and fell to the English. The narrow passage in the rocky cliff where they were let in, can still be seen today. It is, rather appropriately, called the *Porte des Anglais*. In 1471, on the orders of Louis XI, the fortresses were demolished and whatever remained was destroyed, during the Wars of Religion, by the future Good King Henri, Henri of Navarre.

But Saint-Cirq-Lapopie rose again from its ruins, this time to the sound of early lathes, and got on with its day-to-day business of making barrel taps and other wooden objects, providing a living for one thousand five hundred people. At the turn of the century, the *roubinetaïres* – as the tap makers were called, from *robinet*, meaning tap – still worked there, as their fathers had before them, since 1450. Today, a number of craftsmen and artists have revived the tradition and occupy the old workshops that line the narrow streets.

Rising steeply from the fine lancet arch of Rocamadour Gate, Grande Rue is lined with magnificent houses. In the shade of their dark-tiled four-sided roofs and corbelled, half-timbered facades, amid the rich detail of every arch and each mullioned window, the old medieval town seems to come alive again.

Many artists, some very fine and very famous, have loved and lived in Saint-Cirq-Lapopie. There was André Breton, of course, who made the town into a Mecca of surrealism. Then

Saint-Cirq-Lapopie: the house of André Breton

Saint-Cirq-Lapopie: Grande Rue ▶

there was the painter Pierre Daura who found inspiration for his gouaches and watercolours in his retreat, the old 13th century hospital. Paul Paquereau, painter and the most sought-after stage decor designer in the world, also lived here. And Henri Martin painted in a little belvedere perched above the rooftops.
In the 15th century château, La Gardette Castle, the Rignault Museum houses an impressive collection of exotica, with a mixture of African, Pacific Island and Chinese art.
The 15th century church is, of course, fortified. Lending it a certain comforting stoutness, its solid bell tower is buttressed by a turret that encases a spiral staircase. Next to the chancel is a tiny Romanesque chapel where the capital-topped columns are profusely decorated in a charming confusion of interlacing patterns and foliage. Its well proportioned, rib-vaulted nave opens through a double door, which features a central pier and a simple semicircular arched tympanum, with very plain recessed arches.

Past the church, Grande Rue continues its climb up to Place du Sombral, whose pretty name hints at shelter from the sun, and where, accordingly, leafy trees provide cool shade. Further up still, the ruins of the old Lapopie fortress lie at the foot of the rock which was, in former days, the foundation on which the proud keep that once towered here was built.
The tour ends on the rocky promontory of Bancourel, with one last look back over the church, standing tall, the red-brown roofs of the houses tumbling down the length of the slope, bristling with turrets, and the faithful watchtowers, ever on the alert. From here, it is easy to understand how Saint-Cirq-Lapopie came to be one of the first entire villages in France to be classified an historic monument. In the distance, the valley draws the eye down into a paradise of freshness and calm. And, so that not one detail should be omitted, the scent of hollyhocks rises, like some sweet-smelling sacrifice of prayer offered up to the heavens by the immortal soul of these shady recesses.

Saint-Cirq-Lapopie seen from Bancourel Rock

Cénevières

A high cliff rising to a height of nearly seventy metres on the left bank of the Lot River, north of Limogne, a solid rock overlooking one of the entry routes into the region – this is the perfect strategic site of Cénevières Castle. The dukes of Aquitaine had already noticed it and had built a first fortress on this location. Then, through successive additions, the small, simple fort became a composite residence, with its high, 13th century keep and two more comfortable, 15th century main buildings. These were linked together in the 16th century by a colonnaded gallery, very Italianate and Renaissance in style, topped by the regular frieze of fine skylights. Legend has it that Henry of Navarre, the future Henry IV, who besieged and sacked Cahors in 1580, spent several nights with a lady friend in this comfortable residence.

On the ground floor, the room and kitchen are vaulted and the keep room has a trapdoor opening onto the forgotten depths of the cellars. The grand living room on the first floor is decorated with fine Flemish tapestries from the 15th and 16th centuries and a superb painted ceiling from the Renaissance. Also on the first floor, a chamber measuring twenty-five square metres is decorated with thirteen polychrome frescoes showing Ovid's *Metamorphoses*. It was set up at the beginning of the 16th century by Antoine de Gourdon, then the owner of the castle, for his experiments and work in alchemy. In the fireplace, whose hearth was used as the *athanor*, or furnace, by the initiates, there is a reproduction of the philosopher's stone, the mysterious substance whose fabulous powers included the ability to change metals into gold.

Following the river along the same shore, along a road cut into the cliff, we see the village of Calvignac, perched on the high crest above one of the Lot's meanders.

Black Work

The transformation of a base metal, like lead, into gold – such was the quest of alchemists ever since the Middle Ages. In order to proceed with this Great Work, transmutation of matter, the philosopher's stone had to be produced. Another of its properties was the ability to produce an elixir ensuring a long life. This extremely coveted matter was obtained in three steps: the black work or putrefaction, the white work or purification, and finally, the red work, rubification. These alchemists, learned men and philosophers, if they did not receive the self-interested support of certain French kings, such as Louis XI or François I, were often condemned, or even persecuted by the Church, which considered their art to be sorcery. They therefore carried out their work in the greatest secrecy and left little trace of it behind them. Some of them made important collateral discoveries, such as Raymond Lulle of Majorca (1233-1315) with nitric acid and acetone. Their secrets and methods were transcribed in coded or symbolic language. Although they were made during the High Renaissance, the wall paintings of the chamber in Cénevières Castle were inspired by the *Metamorphoses* of Ovid (43BCE-17CE). This was a collection of legends from Greco-Roman mythology, but their themes defined a code of behaviour for alchemists, often shown wearing a Phrygian cap. Hence, on the north wall, Hercules, in his eleventh labour, is stealing three golden apples from the Hesperides Garden away from Atlas, reminding the initiate that telluric treasure is taken by force. Another scene shows Orpheus returning from the Underworld, forgetting Hades' advice and turning around to see Eurydice carried away by the flames. It warns the initiate that passion for gold condemns the Great Work to failure.

Cajarc

With its orderly rows of houses arranged in circles around a little church, Cajarc is a charming little riverside town. The site is very old indeed, even if today there are few ancient remains to be seen. There is, however, still one big residence, the Maison de l'Hébrardie, with pointed arched windows, a reminder that, in the Middle Ages, a fine castle once towered here above the surrounding alluvial plain.
The tiny Chapel of the Capelette, standing on the top of a gently sloping hill, was built during the 12th and 13th centuries. Today, all that remains is its apse, fragile and partly overgrown with ivy. In its day, the bargees came here to pray before embarking on the capricious waters of the Lot.
Upstream from Cajarc, the village of Montbrun clings to its rocky peak, proudly bearing its fine 14th century fortified castle like a stony royal crown.

Montbrun

The Saut de la Mounine

Rocky ridges that follow the curves of a river so closely are always an impressive sight. But a sheer cliff, tens of metres high, plunging vertically towards the perfect arch of a calm river and the complex geometry of patchwork fields, is a picture of such awesome grandeur that it is hard not to be moved by the beauty of such a place. Add legend and drama to this kind of site, and the electrifying result is a place like The Saut de la Mounine – The Monkey's Leap. Legend tells how a sweet and beautiful girl secretly loved the dashing Renault de Salvagnac. But the girl's father, the terrible Lord of Montbrun, with angry shouts and threats, forbade his daughter to marry this neighbouring enemy, and vowed that she should pay the full price of her treacherous love. The punishment was dreadful, but the decision was firm. The girl was to be taken to the top of the cliff and thrown to her death. While the crime was being prepared, everyone was appalled by the father's unyielding hard heartedness. A hermit, who lived in the darker corners of the cliff, took pity on the poor girl and hatched a clever plan. He suggested that the small monkey –called a mounine in Occitan– would be dressed in the girl's clothes and thrown over the cliff in her place. On the appointed day, the monkey, unaware of any mortal danger, let itself be pushed over the cliff and into oblivion. At the sight of his daughter plunging to her death because of his own unthinking arrogance, the father was filled with remorse. Grieving for his daughter, he repented and then, filled with joy at finding her alive, forgave her and consented to her marriage with Renault. And so the site got its name, a reminder of the sacrificial leap made by a monkey – a *mounine* in the local Occitan language. Poor little beast!

Larroque-Toirac

Blending in perfect harmony into its stone setting, Larroque-Toirac Castle consists of a strange collection of buildings that rise in crescendo up to the top of a long thin tower which, wearing a necklace of machicolations, is crowned with a pretty pointed roof.

Owned by the Cardaillac family, the castle was besieged and partly destroyed during the Hundred Years' War. Although disfigured and leveled in part, the castle rose again from its ruins and still has something of the lofty, rather haughty bearing common to all proud historic sites that have "seen it all before".

Close by, the village of Saint-Pierre-Toirac clusters around a highly fortified church-keep. Some of the Romanesque sculptures, like the one of the Angel of Good beating a fabulous creature in single combat, are really magnificent. The sarcophagi near the church are from an ancient Merovingian cemetery.

Saint-Pierre-Toirac: Merovingian sarcophagi

An Underground Museum

South-west of Saint-Pierre-Toirac, in the Aveyron department, the Foissac caves were discovered in 1959. Over eight kilometres, an underground river has dug a series of galleries out of the karstic substratum of the causse. In 1965, the members of the Capdenac Speleological Club found remnants of prehistoric occupation there. These have been displayed on site, making the Foissac caves into a real underground museum. They have been open to the public since 1973. Visitors can admire the beautiful concretions and other fistulous decors of the Obelisk or Rockslide chambers, but they can also see the bones of animals that no longer exist, such as mammoths and cave-dwelling lions, of which Foissac possesses a complete skeleton. The caves were used as shelters for Chalcolithic populations. These famers and herders, who lived here over 4,000 years ago, have left a large, curved jar whose side handles are still visible, scrapers with bone handles, and a granite millstone. Beyond a jumble of rocks, a funerary area on the shores of the underground river hid the skeleton of a woman about fifty years old, a child aged about seven or eight, and an adult man. But the most moving testimony left by these people is the print of a child's small foot, left in some soft clay.

Foissac caves: large pot

Capdenac

Capdenac is an original little town which, for a very long time, has cultivated the fine art of contradiction and contrast. With one foot in each of two *départements* – Lot and Aveyron – Capdenac is subject to the rule of both. But its contrariness does not stop there. The only thing the opposite banks of the Lot have in common here is the beautiful ribbon of water along which they both run. One is gentle and flat, opening onto a little plain where the town of Capdenac-Gare, or Capdenac Station, has grown up at the junction of the rail lines from Rodez and Toulouse. The other rises sheer and abrupt from the river which winds in a perfect curve around a core of hard rock. Clamped to the cliff edge, the town of Capdenac-le-Haut, or High Capdenac, is aptly named.
Archeological digs have found traces of a pre-Roman Gallic *oppidum* on the promontory above the plain. Some even claim that this is the prestigious and much fought over site of the celebrated Uxellodunum, featured in Julius Caesar's writings on the Gallic wars, and the setting of the final tragic episode of Gallic resistance. But his writings say very little of the exact location of the town. The cliff, the river and the spring (known, incidentally, as Caesar's spring), dried up by the Roman troops – all the ingredients are there, but with no real evidence, the theory remains unproven.There is, however, plenty of proof that, in the Middle Ages, this was a fortified town. Even today, it boasts barbicans, lancet arched gates, crenellated keep, bartizans and perimeter walls. A "Hundred Step" staircase leads to the beautiful Roman fountain, hollowed out of the cliff face, the "English Fountain". An essential strategic position, the town was the setting for a long succession of combats and sieges. Henri IV's finance minister, Sully, retired to Capdenac after the death of the King. His residence, extensively restored, still bears the name "Château de Sully".

Capdenac: the English Fountain

◄ Capdenac: the keep

RUE
ORTABADIAL

6

THE CÉLÉ VALLEY

Figeac

On the banks of the Célé, among the last foothills of the mountainous Massif Central, the little town of Figeac stands guard over the age-old routes that descend from the high valleys of the Rouergue region.
The Célé has been forded here since Gallo-Roman times. Sarcophagi and defensive walls dating from this period confirm Figeac's very ancient origins. But the town didn't really start to develop until the year 753 when Pépin le Bref founded a monastery here, as an act of thanks to God for all the easy victories he had been blessed with. When, before his very eyes, flying against a backdrop of blue sky, a pair of doves formed a cross – the most obvious symbol of the Christian faith – the auspices seemed particularly favourable. So Pépin built his monastery here, and Pope Stephen I came to consecrate it in 755.
Little is known about the history of this small religious community between 755 and 838. It probably became sufficiently prosperous to attract the attention of and fall victim to the Vikings, for whom such monasteries, behind whose walls much treasure was to be found, were easy prey. Then, in 838 an abbey was founded here and the little town that grew up in the shadow of its walls began to flourish. One of the stopping points on the pilgrim ways to Santiago de Compostela and Rocamadour, the town gained its independence from the abbey, and was granted many royal privileges, including the very unusual one of striking its own coins. The 13th century *Oustal de la Mounedo*, now known as the Hôtel de la Monnaie, or mint, is a reminder of the importance of this right, even if the building itself was really only an exchange where coins were weighed. Its well restored facade, with a row of gemeled windows above rib-vaulted arcades, a *soleilho* – a traditional open gallery running under the roof – and a lantern opening in the chimney stack, is remarkably elegant.
The little streets running through the old town are very pleasant. The facades of the houses are all decorated with minutely detailed and regularly pointed patterns of small red bricks, framed by the darker pattern of half-timbered crossbars. Right under the edge of the roof, protected by rows of rounded tiles, the open loft, or *soleilho,* fills the upper rooms with warmth and light, sadly lacking in the floors below, making a well aired and well protected all-purpose living and working area.
Dating from the 12th century, the former abbey church, Saint Saviour's, has seen a good many

Figeac: house of Pierre de Cisteron

changes throughout its long history. From its Romanesque foundations to its Gothic vault, its walls bear the record of the different periods of man's past, some brilliant and fruitful, others painful and destructive, some, like the Hundred Years' War and the Wars of Religion, leaving deep scars.

The church was built to a plan typical of this kind of pilgrimage church, designed not just as a place of prayer, but also as a place of rest and refuge for the poor pilgrims treading the paths of forgiveness and mortification. Cutting across the vast nave, the transept forms the cross-shaped plan common to Romanesque churches. An ambulatory, running around the chevet, opens onto a series of apsidiole chapels. Many of the ornately sculptured capitals have disappeared, been moved or even been converted to new uses, such as bases for a pair of stoups, one in each transept arm. Bathed in the soft blue light that falls from its stained-glass windows, the walls of the chapter house are covered in finely worked wooden panels.

A busy market town, medieval Figeac had two open squares, squeezed in among its tangle of narrow streets: Place Carnot, formerly known as the Lower Square, and Place Champollion, formerly known as the Upper Square. Although the wheat market was destroyed in the 19th century, Place Carnot still boasts some fine houses, like that of Pierre de Cisteron, which, with its corbelled turret, is a reminder of the success of this clever armourer who, although thoroughly Protestant, was nevertheless supplier to His Very Catholic Majesty, King Louis XIV.

The facades around the Upper Square, like that of the former house of the commander of the Order of the Knights Templars, open through finely decorated vaulted windows.

Nearby, the birthplace of Jean-François Champollion, the very learned Egyptologist, houses the Champollion Museum, which is dedicated to his work. Its collections include a number of fine Egyptian antiquities and, in particular, one of the rare moldings of the famous Rosetta

Figeac, St. Saviour's Church: a carved wooden panel

The First Egyptologist

Born in Figeac in 1790, the young Jean-François Champollion, enthralled by ancient writing, was to dedicate his short life – he died in Paris in 1832, at the age of 42 – to deciphering the mysterious signs decorating Egyptian monuments, hieroglyphics, those strange "sacred signs". He came into the possession of a copy of the Rosetta stone, a fragment of black basalt, a metre high and seventy centimetres wide, on which a text was written in three different writings – Greek, demotic and hieroglyphic. The original stone, rapidly confiscated by the British and placed in the British Museum, had been discovered at Rosetta in1799, during Napoleon's expedition to Egypt. One of Napoleon's generals had translated the Greek inscription. It mentioned a priestly decree, promulgated in 196 BCE, setting the honours to be rendered to Ptolemy V in Egyptian temples. Champollion identified the signs that composed the words. Beside the phonogrammatical signs, corresponding to sounds, the hieroglyphs contained signs regrouping several consonants and ideogrammatical signs where images provided the meaning of words.

"Place of Writing" Ex-Libris Jean-François Champollion by Joseph Kossuth, Champollion Museum, Figeac

After a communication to the Academy of Inscriptions and Literature in 1822, Champollion published in 1824 his *Précis de système hiéroglyphique* in which he disclosed the secrets of this writing that for so long had been relegated to a beautiful stylized decoration, since no-one had been capable of reading it. In 1826, he was named curator of the Egyptian department at the Louvre Museum where, in 1831, he inaugurated new teaching at the Collège de France, where a chair in Egyptology had been created especially for him. Champollion wrote two books, published after his death by his brother, Jacques, an Egyptian *Grammar* and a *Dictionary*, that established the bases for scientific research in Egyptology.

Cast of the Rosetta Stone, Champollion Museum, Figeac

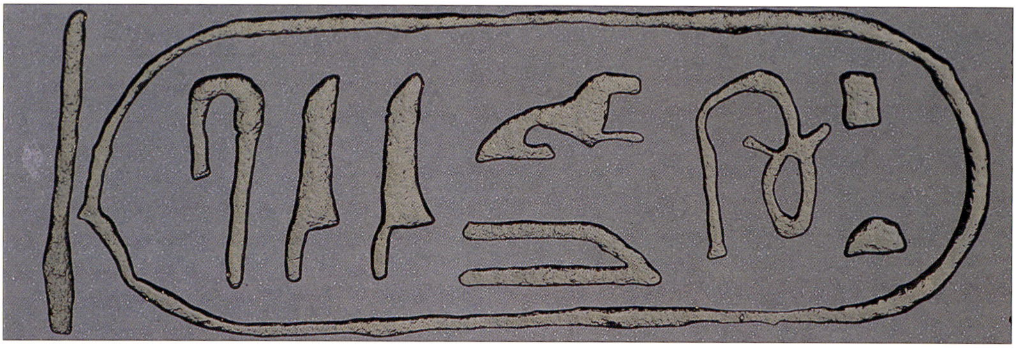

A Church Within the Church

He Benedictine Abbey of Cluny was founded in 909 or 910 by Guillaume Ist, Duke of Aquitaine, in the Mâconnais area. Placed under the authority of the pope, by 931 it had become the centre of a monastic reform that spread throughout Europe. Some reformed abbeys, like Rocamadour, remained independent. Others, keeping their status or by becoming priories, joined the *Ecclesia cluniacensis*, the Cluny church that included over a thousand establishments at the beginning of the 12th century. Most of them were priories, whose prior was named by the abbot of Cluny, such as Souvigny, Sauxillanges, La Charité-sur-Loire, Lewes and Saint Martin in the Fields – the "five daughters" of Cluny. But the *Ecclesia* also included obedient abbeys, such as Moissac or Beaulieu-sur-Dordogne, whose abbots were freely elected but swore fidelity to the abbot of Cluny. During the 12th century, the *Ecclesia* became a structured order. Its establishments were spread over ten provinces: Auvergne, France, Gascony – to which Figeac was attached, Cluny, Provence, Poitou, Germany, Spain, England and Lombardy. Wearing a homespun monk's habit, the Cluny "black monk" spent most of his time at prayer, to the detriment of physical work dear to Saint Benedict. The Cluny monks played a essential role in the Christianity of the 10th, 11th and 12th centuries, supporting the Peace of God and the Gregorian reformation, participating in the rise of the Compostela pilgrimage. Their richly decorated sanctuaries, such as Cluny III, reached grandiose proportions. But in the 11th century, new monastic orders such as Cîteaux, preached a return to Benedictine ideals. After the Abbot Peter the Venerable (1122-1156), the rise of Cluny ended. But until the beginning of the 14th century, Cluny retained power and prestige before its decline started.

stone, which provided the first key to deciphering the ancient hieroglyphic symbols of sacred Egyptian writings.

Perched on a small hill, one of Figeac's oldest buildings, the church of Notre-Dame-du-Puy, Our Lady of the Mount, enjoys a commanding view. Its Gothic nave was considerably altered in the 17th century, but its apsidioles still have some fine Romanesque capitals. In the chancel, the altar boasts a large walnut retable. The significance or the purpose of the pair of long stone needles that rise on the hilltops of Lissac to the west and Cingle to the south remains a mystery. Standing seven to eight metres above their pedestals, narrow and pointed, and bearing neither inscription nor mark, these monuments have stood like silent sentinels since the 12th or 13th century. Perhaps they once indicated the limits of the monastery's domain, or marked the great pilgrim ways of the past, or perhaps they were built as some kind of mysterious votive offering. It seems that nobody will ever know.

Figeac: Lissac needle

Canons and Monks on the Shores of the Célé

In the valley of the Célé River, where today sits the little village of Espagnac-Sainte-Eulalie, in the 12th century Bertrand de Griffeuille founded a priory of canons who had adopted the rule of Saint Augustine. In 1212, the canons were replaced by canonesses of the same obedience. The priory benefited from the generosity of the Hébrard de Saint Sulpice family to whom the valley of the Célé belonged as a fief. In particular, of Aymeric Hébrard de Saint Sulpice, Bishop of Coimbra, who, in1283, had the priory moved from its often flooded location. The priory then took the name of Notre-Dame-du-Val-Paradis. Devastated during the Hundred Years War, the priory was rebuilt in the 15th century and was finally

Marcilhac-sur-Célé

Espagnac-Sainte-Eulalie: the priory

abandoned only at the Revolution. The priory building with its fine soaring steeple dates from the 15th century. From the ancient 13th century church, only the walls of the nave, a gateway and the ruins of bays remain. Inside, near the chancel, three recumbent statues are perhaps those of the tombs of the knight Hugues de Cardaillac, wearing a surcoat and helmet, with sword and shield, his wife, Bernarde de Trian and the bishop Aymeric Hébrard. Upriver, in the heart of a cirque of white faced cliffs, the village of Marcilhac-sur-Célé is dominated by the imposing mass of its former Benedictine abbey. In the 11th century, this abbey owned a small sanctuary that it rather forgot about. It was Rocamadour. The monks of Tulle took advantage of this to move in. In 1166, the body of Saint Amadour was discovered and Rocamadour became the centre of an important pilgrimage. The monks of Marcilhac then tried to recover their sanctuary and the Tulle monks were chased out. Lawsuit followed lawsuit. The intervention of the highest ecclesiastical authorities – the bishops of Cahors, of Bourges, and the pope himself -- was needed to settle the quarrel. Finally, the monks of Marcilhac agreed to abandon their rights over Rocamadour in return for payment in cash. The abbey was prosperous but the Hundred Years War dealt a sharp blow to the small community, which was later weakened further by violent disagreements during the Wars of Religion. The Romanesque section of the abbey church is in ruins and is open to the skies. Only a large square tower is well preserved and it overlooks the high walls covered with ivy and columns with capitals.
The southern doorway is decorated with a tympanum sculpted with the Last Judgement. The Gothic section, rebuilt in the 15th century in Flamboyant style, contains panelling, frescoes and fine armorial coats of arms. The 12th century chapter house, is decorated with historiated capitals where fabulous animals, half-monsters and half-demons, show terrible cruelty in tormenting the poor sinners lost in Hell. Some figures with celestial smiles contemplate, in a vision of ecstasy, the Paradise they have finally attained.

Christ in majesty, teaching, capital in capitular room, Marcilhac-sur-Célé Abbey

Cabrerets

The village of Cabrerets nestles at the foot of the high cliff of Rochecourbe, where the Célé and the little Sagne meet. Clinging to the rocky face, the ruined walls of the Château du Diable, or Devil's Castle – also known as the Castle of the English – blend into the natural stone wall of the huge cliff. Protecting the troglodytic settlement, a huge edifice rises on the edge of the cliff. Flanked by a big tower, all that can be seen of this building is its immense facade which, apart from a single high mullioned window – obviously a look-out position – is entirely blind.
Lost somewhere in the first misty centuries of the Middle Ages, the origins of Cabrerets are uncertain. However, its presence and the dread it inspired are both clearly recorded during the Hundred Years' War, when roaming bands of looting mercenaries, fearing neither God nor man – *the Grandes Compagnies* – set up their headquarters there. Ruling with violence and terror from this impregnable stronghold, they gave Cabrerets a sinister, almost something of an evil, reputation.
Downstream from the village, Gontaut-Biron Castle, built in the 14th and 15th centuries, towers above the valley. Its main building, of golden stone, is flanked by an enormous round tower with a conical roof. A covered way leading to the top of its high walls throws a soft corbelled shadow onto its light-coloured facade. Set firmly on solid stone arches, a beautiful terrace, surrounded by an ornamental balustrade, overhangs the road. At the top of a little rise is Cabrerets' Romanesque church, extensively restored in the 19th century.
Three kilometres from Cabrerets, an underground river emerges from the Gramat limestone plateau as the Pescalerie Spring. Its dark waters seem to hold back a moment until, struck by sunlight and gently babbling, they escape into a small pool, then cascade noisily down into the valley.

Pescalerie Fountain

The Quercy Open Air Museum

The Quercy Research Association, founded in 1974 by J-L Obereiner, approaches all its themes with relevance and seriousness, be they historical, geographical, sociological, cultural or economical, when it deals with this beautiful region. The idea of heritage, taken in its broadest sense, was obviously not missing from the group's investigations when, in 1985, decided to open to the public the Quercy Open Air Museum (musée de Plein Air), around the ruins of old Cuzals castle. Even though it is partially destroyed, the building still retains a lot of elegance with its two square towers framing a facade with wide bays opening onto the blue sky.
A large staircase climbs towards the monumental entrance crowned with a fine balustrade. Around the castle, its fifty-hectare domain is filled with evidence of traditional Quercy architecture: dry-stone *caselles*, stable-barns, and pigeon lofts, but also faithful reconstructions of old farms presented in parallel for two very different periods – just before the French Revolution, and the very beginning of the twentieth century. The Ancien Regime farm is a small building with thin walls, reinforced with simple half-timbering, topped with a large thatched roof.
A plain wooden partition separated the stable from the single living room. Few openings kept out the cold, but also light. A crude stone sink led dirty water towards the pig-sty. The interior furnishings were summary – a few chests, a table with a long wooden bench on either side, a rudimentary bed on which a soft eiderdown was often replaced by straw, several pottery or wickerwork objects, and the indispensable tool – a sharp knife. These were all the furnishings the

Bread oven, early 20th century

A Clayton touring locomobile

inhabitants possessed. By the beginning of the 20th century, the house was spruced up. It was of fine causse stone, with a garland of grapevine running just below the roof. The animals had completely left the main dwelling for a stable-barn, built away from the house. Many more utensils and tools had become available and there was even a kind of kitchen – la *souillarde* – next to the main room with its fireplace, the *cantou*. This was so deep that one could coil up in it, keeping warm at the comfort of a hearth, in which a fire remained lit all the time. The table is long and solid, polished to a gloss over the years. Against the whitewashed wall, an old clock still marks the passage of time. Tousled corncobs hang from the ceiling, along with a huge ham, and round loaves of bread, aligned in their rack, give off the delicious smell of fresh bread, like a forgotten memory. Maintaining ancient know-how and techniques no longer used is one of the purposes of this local ethnology centre, since it is certain that a region's heritage is not simply confined to its architecture, even though this may be superb. The clog-maker, potter and blacksmith still work before our eyes. Out of his beautiful oven, the baker removes the balls of dough that have risen on their *paillassous*, wicker baskets that were used as baking pans. The extraordinary combine-harvesters associated with steam locomobiles are still in working condition and are used for very impressive demonstrations. The museums of Water, Fire, Pottery and Copperware, Basket-weaving, Tools, Bee-keeping, and Rural Crafts are all extremely rich. Finally, in a humorous wink at fictional history, the museum presents an excavation site for archaeologists of the future who have come to dig up the bizarre remnants of a very ancient civilization dating from the end of the twentieth century CE.

Excavation of a late 20th century site by future archeologists

Pech Merle Cave

Pech Merle Cave, three kilometres from Cabrerets, was discovered in 1922 by two fourteen-year-old boys, André David and Henri Dutertre, who had participated in exploration undertaken by Father Lemozi, the priest of Cabrerets.Open to the public since 1924, the cave extends over 1,200 metres of galleries dug over several levels. It is rich in concretions of all kinds and contains that very rare thing – *cave pearls*. But what makes this site exceptional are the paintings and engravings on the walls. They date from the Final Perigordian period to the Middle Magdalenian. There is a total of nearly five hundred designs.

In the *Mammoth Chapel*, a large black frieze has been made, starting from a centrally placed horse, surrounded by a circle of mammoths, bison, horses and aurochs. Along with Bernifal and Rouffignac in Dordogne, Pech

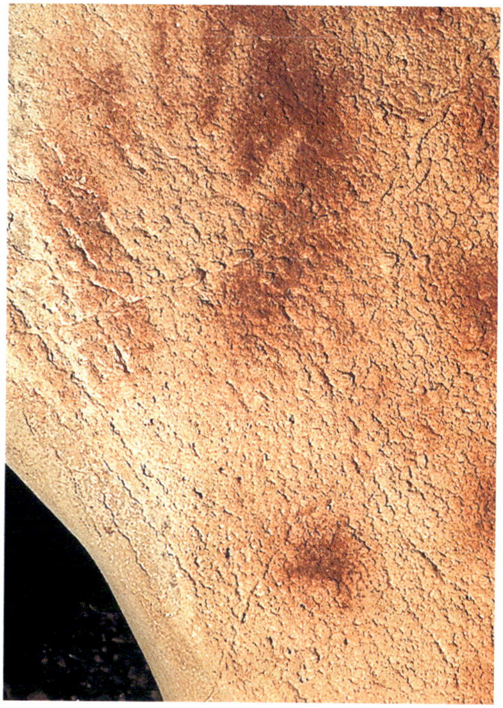

Painting chamber: negative hand and red dots

Black frieze, also called Mammoth Chapel

◄ Pech Merle Cave, painting chamber: dappled horse (detail)

Merle is one of the rare decorated caves in which mammoths are shown in large numbers. The ceiling of the *Hieroglyphic Chamber* is covered with signs and drawings done with the finger, among which mammoths, ibex, and heavy-breasted female figures can be made out.

In a place where water used to collect in the *Disc Chamber*, footprints made by a child and a woman have been left in the clay. But the most spectacular wall painting in the cave is the great fresco of dappled horses. Starting at the characteristic shape of a bump in the rock that clearly looks like a horse's head, the artist ended the drawing with two horses leaning back, covered with big black spots. Their small heads, powerful bodies and thin legs are a reminder of Lascaux. Six hands, with their outlines in a negative on the rock-face, surround the composition. The fine red outline of a long fish adds even more mystery to this strange scene. In another chamber there are bones of cave bears.

Pech Merle Cave: adolescent's footprint

The Lemozi Museum

The Pech Merle Prehistory Centre obviously includes Pech Merle Cave along with the Amédée Lemozi Museum, which opened to the public on April 5, 1981. It bears the name of the man who made Pech Merle cave known and who founded the first museum in Cabrerets. Its vocation as a regional prehistory museum is clearly defined by those in charge. This museum, which presents very fine collections of pieces discovered in the one hundred and sixty deposits in Quercy, sums up the current situation of research undertaken and underlines the originality of this region that is so rich in ancient remains. From the most distant Paleolithic to the Neolithic, all the periods of this long era preceding our history are evoked here. Objects discovered near dolmen, so numerous in Quercy, like superb silex Chalcolithic daggers, the beautiful ornaments, axes, swords and all the ceramics from the Middle Bronze Age, make it an exceptionally rich museum. Well integrated into the Pech Merle site, its purpose is to prepare the visitor for a visit to the cave. Its two exhibition rooms, a large projection room, its library, archive room and storeroom give it the well-deserved title of the Quercy Prehistory Museum.

Ceramic pot from Middle Bronze Age

7

FROM THE AVEYRON GORGES TO WHITE QUERCY

Beauregard

Like all the bastides created during the Middle Ages, Beauregard has retained its checkerboard layout, with its straight streets meeting at right angles. Founded by the abbot of Marcilhac, it was subsequently purchased by the King during the 14th century.

On the central square where the heart of the town used to beat, a little market hall was built in the 14th century, evidence of the commercial activity of this town that concentrated, within its fortified walls, the riches and production from the surrounding countryside. The hall is covered with a heavy flat-stone roof, laid in layers on the four sides of the pyramid-shaped beams. Short stone pillars hold up the incredible weight of the roof. The memory of small, noisy, colourful and busy markets is suddenly conjured up, if we take the time to come close and to admire the fine grain measures that have been retained. Carved in light, blonde stone, shiny and polished by the thousands of merchants' gestures, and the pouring of the grain through the necks, these measures still stand here, apparently empty and useless, but still measuring the tenuous weight of the past. The 14th century church was much restored during the 19th century, but still protects the relics of Saint John. An old door with an emblazoned lintel is becoming overgrown by persistent vegetation.

Dolmen and menhirs are numerous in the surroundings, like that of La Borie du Bois.

Dolmen at La Borie du Bois

Laramière

The little village of Laramière, with its collection of old 12th century priory buildings, is set in the middle of the Limogne plateau, on the pilgrims' way to Santiago de Compostela. The little priory was one of many, founded by the Augustinian order and dependent on the great Abbey de la Couronne in Charente, that were dotted along this route by which, in times past, pilgrims traveled on their way to Spain and the shrine of Saint James in Compostela, hoping for miracles of healing, or forgiveness, or to earn themselves a place in paradise.
But the priory was not always a place of refuge and a haven of peaceful contemplation for pilgrims concerned only with the salvation of their souls. The 16th century was a time of violent confrontation between Catholics and Protestants and the little priory, caught up in a series of bloody fratricidal wars, was looted and suffered terrible damage time and time again. But when, in the 17th century, the Jesuits decided to repair the priory buildings, and notably the Bailiff's house, which was enlarged and restored, the priory was saved from certain and permanent demise.
The priory has a fine bell gable, supported by sturdy buttresses, with three arches for its bells. Throughout the Middle Ages, the priory was a place of refuge for pilgrims on their way to Compostela, and its great hall served as hospital, hospice and hostel. The door leading from chapter house to chapel has two beautiful rib-vaulted bays. Their finely sculptured, elegant arches rest on a pair of columns with capitals chiseled into the likenesses of the very pious Saint Louis and his unyieldingly determined mother, Blanche of Castile. Even if they hold their heads rather stiffly and their hair is a little too perfectly curled, the respectful homage paid here to mother and son is a reminder that both were ardent defenders of Catholicism. The walls and vaulted ceilings around them are painted in geometric designs.

Laramière: chapter house in priory

◄ Laramière, priory chapter house: capital showing Saint Louis

Caylus

From its vantage point on the ridge of a small hill that rises just on the edge of the limestone plateau, Caylus has, for many long centuries, stood faithful watch over the intersection of two ancient Roman roads.
During the 12th century, a fortress was built at Caylus, and a huge keep was added in the 13th century. Today, however, all that remains are a few ruined walls. One of the largest noble domains in Quercy, Caylus was to witness the arrival of Simon de Montfort's troops during the Albigensian Crusade. Although restored to Catholicism, Caylus had not seen the last of the violence bred by outraged faith. Sacked and burnt in the 16th century, during the Wars of Religion, in 1622 Caylus was established as the headquarters of Louis XIII who, along with his faithful Cardinal Richelieu, had decided to finish the Huguenots off by besieging their strongholds, such as Saint-Antonin-Noble-Val. Caylus still has the narrow streets typical of medieval times, lined with fine half-timbered, corbelled houses. In Rue Droite, The Maison des Loups – the Wolves' House – has a number of decorative pointed arches and small elongated windows. It gets its name, of course, from the four wolf-shaped gargoyles, alert and ready to leap, that cling to stone supports below the roof. The houses that surround Place du Marché – the market square – open onto the arcades of a covered walkway. The town's old grain measures can still be seen on one of the low walls in the old market hall. Made of golden stone, and smoothed and polished by time, they have a certain, undeniable, charm.
Saint John the Baptist's Church was once a fortified place of refuge, as its solid buttresses and thick high walls suggest. In spite of its mainly defensive function, the graceful shape of its attractive bell tower gives it a certain elegance. Circled by side chapels, the single 14th century nave leads into an exceptionally beautiful chancel. As if held in an ethereal whirl-

Caylus: grain measures at old market hall

◀ Caylus: Wolf House

wind, twirling rays of sunlight catch the dazzling colours of the stained-glass windows, like a thousand facets on a giant kaleidoscope. To the left of the chancel, a remarkable figure of Christ, sculpted by Zadkine, conveys, in the contortions of his tortured, emaciated body, all the pain of his terrible agony. The work is particularly astonishing since it suggests far more than it actually shows. Only the gesture of a hand clutching at a nail recalls the crucifixion. The absence of the cross only increases the sense of suffering, expressed simply in the rigid movement of an outstretched arm and its shadows.

The porch is decorated with sculptures of a naïve style, illustrating the popular themes of the crucifixion and a pietà, where the figures are very stylised and their forms extremely simplified. One pillar bears a very strange scene. A fat face, wearing a broad grin that reveals two rows of regular teeth, has been chiseled out of the stone capital. A fantastic beast, a heavy-clawed dragon, perches on his forehead and, with no consideration for his fat host, leans on his nose. To one side, a small, long-eared animal grips a man's head, his hair in ringlets, in its talons. The perspective, gestures and bizarre forms of the figures all suggest the horrifying vision of the world hereafter that seemed to haunt the imagination of so many good souls of that time.

The little pilgrimage chapel Notre-Dame-de-Livron – Our Lady of Livron – was built next to a spring of clear water. Legend tells how a dreadful dragon once lived here, terrorising the whole region. With a little help from Our Lady, an exceptionally courageous knight is said to have delivered the local population from its undesirable neighbour. The chapel was built here in the 13th century and, a little later, a splendid Gothic chancel was added. A few kilometres from there, and well worth the detour, another pilgrimage chapel, Notre-Dame-des-Grâces – Our Lady of Mercy – stands on the top of a promontory and, with its fine carved doorway, is a picture of elegance.

Caylus: porch of St. John the Baptist Church: Virgin with child

Notre-Dame-des-Grâces Chapel: the doorway

Beaulieu en Rouergue

Set in the lovely valley of the Seye River, a calm tributary of the Aveyron, Beaulieu en Rouergue Abbey is a place of such great beauty, that the first glimpse of it causes a little tingle of pleasure and excitement. The abbey buildings are laid out to a plan typical of a Cistercian abbey. Around the quadrangle where the cloisters once stood, the church, chapter house, monks' dormitory, refectory and storeroom still remain, built to be functional and efficient, in keeping with the austere hearts and minds of the order of monks from Cîteaux who lived here.

Beaulieu was founded in 1144 by Adhémar III, Bishop of Rodez, as a dependent of Clairvaux Abbey. Its name "Beaulieu" – or Belloc as it was called in the Occitan language – literally means "beautiful place". The abbey rode out the violent storms of the Middle Ages with surprisingly little damage. Throughout the dreadful repression of the Cathar heresy, the interminably long Hundred Years' War, and the atrocious settling of doctrinal scores in the Wars of Religion, only the ethereally fragile cloisters were lost. During the Revolution, the abbey was sold to the state as a national monument and was converted into a farm. Paradoxically, this was to be its salvation. The fabric of the church-stable was maintained, the storeroom continued to be used, and the abbey was saved from abandon and certain ruin.

The church's decor is extremely bare and simple. Lit by long bays, a single nave opens at the transept crossing into an octagonal dome, set on a lovely squat bell tower. Its magnificent rose windows are cut into symmetrical arabesques with the figure seven in leitmotiv. Apart from a little paschal lamb suspended from the apsidiole vault, there is no figurative decoration. Opening onto the place where the cloisters once stood, the chapter house, with pointed arches resting on two heavy pillars, contains a sober stone bench.

The Cistercian order

In 1098, Robert, Abbot of Molesmes monastery in Burgundy, left his community accompanied by several monks who wished, like him, to follow the Benedictine rules in their original strictness: living in poverty, spending their time in prayer and manual work. They founded the Novum monasterium on the holding ceded to them by the Viscount of Beaune, south of Dijon. It was a wooded and swampy place called Cistels. The beginnings were difficult but under the leadership of Étienne Harding, starting in 1111 on, the monastery developed and was reinforced by new members. Hence, in 1112, Bernard, son of the lord of Fontaine, arrived at the abbey. The number of brothers at the New Monastery, which took the name of Cîteaux in 1119, allowed it to spread and new abbeys were started: La Ferté in 1113, Pontigny in 1114, Clairvaux – where Bernard became abbot – and Morimond in 1115. These multiplied in their turn and Morimond produced Aiguebelle or L'Escale-Dieu, which in turn founded several abbeys in Spain. The development of the Cistercian order, based on the texts of the Charte primitive in 1114 and the Charte de Charité of 1119 was considerable and was catalyzed by the exceptional personality of Bernard de Clairvaux. He became one of the leading figures of Western Christianity. Very attached to his order, he made sure that its development and enrichment did not alter the demands of simplicity and severity wanted by its founders. This requirement is particularly visible in Cistercian architecture which is marked by austerity. At the death of the man who was to become Saint Bernard in 1153, the order had 350 abbeys distributed all over Europe. In 1250, there were 650.

Saint-Antonin-Noble-Val

Where the Aveyron meets the Bonnette, the valley spreads out beneath the imposing cliff face of Rochers d'Anglars – Anglars Rocks. The site is very picturesque, and the valley so beautiful that, long ago, it earned its old Latin name *Nobilis Vallis*, meaning "noble valley". With the spread of Christianity, the town was dedicated to Saint Antoninus, and added his name to its own. Legend tells how, guided by the hand of God, a small boat carrying the relics of this saint ran aground on this very spot. Reality or fantasy? Nobody will ever know. Real enough, however, was the Benedictine abbey founded here, and whose buildings once stood on the banks of the Aveyron, even if today nothing remains of it.

Saint-Antonin-Noble-Val, so well placed at the meeting point of two valleys and on the borders of three provinces – Quercy, Rouergue, and Albigeois – has had a very eventful

Museum of Old Saint-Antonin: municipal coat of arms

Saint-Antonin-Noble-Val seen from the cliff of Anglars Rock

◄ Saint-Antonin-Noble-Val: former town hall

past indeed. On the edge of Rouergue, which became an English possession during the Hundred Years' War, after the Brétigny Treaty in 1360, the town changed overlord and master time after time, tossed on the changing tides of military fortunes of these secular struggles. But at last, peace returned, and the leather and cloth merchants began to prosper, building beautiful houses, like the House of the Amours, with extravagant and delightfully ornate facades, opening through mullioned windows onto narrow streets that run up the banks of the Aveyron to the central covered market. Near to the market place is the former town hall, a vast 12th century Romanesque house. Over wide arcades on the ground floor, a regular frieze of small columns with sculptured capitals runs the width of the facade. On one pillar, Adam and Eve, in the shade of the fateful tree, prepare to bite the sweet fruit of their bitter crime. On the other, the learned Emperor Justinian holds a large open book, representing his famous code of civil law.

Facade of former town hall: Adam and Eve (detail)

This period of peace and prosperity, however, was not to last, and the busy little town was soon overtaken by a rising tide of murderous folly as religious men, all firmly convinced of the divine justice of their cause, battled against each other in the Wars of Religion. Adhering to the doctrines of the "reformed religion", Saint-Antonin-Noble-Val suffered terribly as Louis XIII and his faithful Richelieu fought the Huguenots to obtain the surrender of the Protestant strongholds that had been granted – rather too generously in their opinion – by Henri IV's famous Edict of Nantes in 1598. The bitter memory of the royal defeat at Montauban and Nègrepelisse in 1621 was still very fresh in their minds when, the following year, the Constable of Luynes, a fine strategist and a clever psychologist, suggested that the King might have a better chance of success if he attacked Saint-Antonin-Noble-Val, which, although a smaller town, was no less defiant of his royal authority. On the site of the little chapel of Pech Dax, where the Calvary cross rises above the town to the west, Louis XIII's military council advised the King on how best to lay siege to the town that lay at their feet. For the French King, whose Catholic reconquering crusade was, in 1622, in full swing, and who needed to wipe out the humiliating defeats of his previous campaign, the success of his operation was of vital importance. And the operation was, indeed, a success. Nègrepelisse was sacked and Saint-Antonin-Noble-Val invaded by royal troops.

Bosc Cave, a small cave hollowed out by a river which dried up long ago, boasts both dripstone formations and prehistoric remains.

Further to the north, Cas Castle rises above the Bonnette valley.

Upstream, on the right bank of the Aveyron, Varen lies in the shadow of the two heavily fortified buildings that have marked its history – the Deanery and the Romanesque church. Downstream, set solidly on the edge of the plateau, the village of Brousses towers above the river from the top of its sheer rocky cliff.

Penne-d'Albigeois

Daring, boldness, and a fine sense of balance. Gazing upwards from the foot of Penne Castle, these are the words that spring to mind. Built on a bare peak, eroded to a stump of hard rock, the fortress – the astonishing handiwork of intrepid man – extends the line of the natural rocky face skywards. Today, all that remains are crumbling towers, ruined walls and a nostalgic yearning for a more heroic age, in centuries past, which was capable of building such aerial retreats.
The village of Penne d'Albigeois clings to the bare sides of the rocky peak. It was, in earlier times, surrounded by walls. As reminders of these days, its fortified gates, reached by a long zigzagging slope, still open at the bottom of the hill. Its church was fortified as well, playing an essential part in defending the village and watching over what was once a major route through Quercy.

Penne-d'Albigeois: the castle seen from the village

An Untameable Race

On April 12 1229, the Treaty of Paris put an end to crusades started twenty years earlier on the lands of the South of France by the pope and then by the King of France. A cornered Raymond VII signed it and in doing so capitulated to Blanche of Castile and Louis IX, accepting the eventual annexation of the County of Toulouse by the Kingdom of France. The treaty's article 31, the longest of the 32 it consisted of, was partly dedicated to the single castle of Penne d'Albigeois. Raymond VII agreed to the following text: "I will deliver to the King the castle of Penne-d'Albigeois by August 1st, and if I cannot hand it over by that time, I will besiege it and will not cease making war on those who occupy it until I receive its submission." The chroniclers of the Albigensian Crusade, although very precise, barely mention this castle, the seat of a diaconate of Albi's Cathare church, but enough to be understood that this eagle's nest was that of proud lords. In 1213, one of them, Olivier, had paid homage to Raymond VI, Raymond VII's father. He had thus clearly chosen the Occitanian camp against that of Simon de Montfort, the crusade's military chief. Shortly afterwards, the latter's brother had besieged the neighbouring castle of Puycelsi. He had raised the siege when the lords of Puycelsi promised to surrender in case Penne gave up or was taken, but de Montfort carefully avoided provoking Penne. Thirty years later, or fourteen years after the Treaty of Paris, Raymond VII wrote to Louis IX that he had been unable to hand over Penne because its lords refused to obey him. In 1248, the Count of Toulouse's agreement to leave for the Holy Land was rewarded by the French King's restitution of Penne-d'Albigeois. At that time, the eagles of Penne had folded their wings.

Bruniquel

Coming out of its gorges, the Aveyron River receives the waters of its tributary, the Vère. Set on a steep cliff, Bruniquel Castle overlooks the confluence of the two rivers. The oldest remains date from the 12th century but Grégoire de Tours, in his *Histoire des Francs*, wrote that Queen Brunehaut had a first fortress built during the 6th century. The Counts of Toulouse came into possession of this castle and had its defences reinforced. In 1211, Raymond VI took shelter at Bruniquel. The situation was difficult as Simon de Montfort conquered, one by one, the Toulouse count's fief holdings in the nearby Albigeois area. Raymond VI's half-brother, Baudoin, who had joined the crusade, intervened and suggested to the knights of Bruniquel that they transfer to him the homage that they paid to his brother. Curiously, Raymond VI accepted and took shelter in Toulouse. Bruniquel became the fief of Baudoin, who welcomed there Guillaume de Tudèle, the author of the first part of the *Canso*, a chronicle in verse of the crusade against the Albigensian heresy. But the proud Saint-Gilles family would not forget the affront. In 1212, mercenaries paid by Raymond kidnapped Baudoin in Montcuq and imprisoned him in Montauban. Raymond VI had his brother hanged above the shores of the Tarn River. Bruniquel is spread over the slopes of the rocky peak. Corbelled houses with mullioned windows squeeze tightly along narrow streets that climb towards the top of the cliff, topped by a keep or Queen Brunehaut's Tower, the New castle and the Old castle preceded by a barbican and a line of ramparts. Next to them, the Knights' Room, from the 12th and 13th centuries, is decorated with thin columns topped with capitals. On the first floor of the New castle, the weapons room contains a 17th century carved wooden fireplace. The Old castle's central building dominates the cliff and opens onto the landscape of the valley through a Renaissance-style gallery dating from the 19th century.

One Against Many

Legend has it that she was disturbingly beautiful and of superior intelligence – formidable qualities when they are used to further great ambition. Born in the Iberian peninsula in about 543, Brunehaut was the daughter of the Visigoth King Athanagild. In 567, she was married to Sigbert I, King of Austrasia, who was assassinated in 575 at the instigation of Fredegonde, the Queen of Neustria. Her son Childebert being a minor, Brunehaut became the Regent of Austrasia until she was chased away by a rival faction and came to seek refuge in Burgundy. The Frankish kingdom had been divided since the death of Clovis in 511 and was torn apart by intrigues. Gontran, the second son of Clotaire Ist and King of Burgundy, wanted to restore some order in his grandfather's mishandled inheritance and organized his own succession by the Treaty of Andelot in 587. Brunehaut then asked to recover the Bordelais, Bigorre, the Béarn, Limousin and Quercy, which had been bequeathed to her by her sister, Queen Galswinthe. But she obtained only Quercy where she perhaps stayed in the castle overlooking the confluence of the Aveyron and Vère Rivers, the *Castrum Brunichildis*, that became Bruniquel. In 597, at the death of her great rival Fredegonde, Brunehaut tried to extend her power over the Frankish kingdom that she wanted to unify. But she came up against the strong opposition of the Frankish aristocracy united around Clotaire II, the son of Fredegonde and King of Neustria since 584. Clotaire II captured Brunehaut, inflicted great violence upon her and finished her off by attaching her to the tail of a horse set to gallop away. The mutilated remains of the beautiful Frankish princess were burnt so that nothing would be left of her. The legend remains. The cultivated Brunehaut, filled with the brilliant Roman heritage that the Visigoths had been capable of adopting, was not accepted by the overly crude Franks.

◄ Bruniquel: Vère River and the castle

Montricoux

Montricoux sits atop a rise on the left bank of the Aveyron River. Dominated by the keep of the former Templar commandery and by the church steeple, the village spreads as far as the river bank, beyond its former fortifications. The Gothic church, named for Saint Peter, contains several 17th century reredos and baptismal fonts of pink marble from... Montricoux!

To the south-east, in the Vère River valley, at the edge of Grésigne Forest, Puycelsi Castle was besieged in 1213 by Guy de Montfort, the brother of Simon, during the Albigensian Crusade. The region was receptive to heretical ideas, and over a century later Monclar-de-Quercy, south of Montricoux, still sheltered a few of the last Good Men of the Cathares. Captured in 1306, they were interrogated by the Dominican Bernard Gui, in charge of the Inquisition tribunal in Toulouse.

Montricoux: the square keep

Like a Prayer

Marcel Lenoir was born in Montauban in 1872. At the age of seventeen, his talent for drawing, as well as his know-how in the chiselling of jewellery, made him leave for Paris. As a student at the school of Decorative Arts, he spent much time in museums, particularly that of the Middle Ages at the Hôtel de Cluny, where he was enthralled by the work of the illuminators, and the Louvre Museum where he took his time contemplating the work of Botticelli and Fra Angelico. He made a few illuminations, became interested in Pointilism – his Coronation of Christ (1898) is a reminder of Seurat –, frequented the Symbolist poets, became initiated into the esoteric world of the Rosicrucians. Meanwhile, his studio in the rue Notre-Dame-des-Champs received visits from Modigliani, Fernand Léger, Bourdelle and Picasso. During his visits home, he painted landscapes and, in 1912, he learned the technique of making frescoes that he put into practice in Bruniquel and at the Catholic Institute in Toulouse. Nor could he avoid becoming interested in Cubism. In 1915, Cézanne's influence began to appear in his paintings. Starting in 1916, he began to use the Art Deco style. In 1931, he died in Montricoux, where Mr. Claude Namy has gathered together, in the former Templar Commandery, a hundred and twenty of Marcel Lenoir's works. He was a genial, eclectic and independent artist who often said, "My prayer is painting."

Marcel Lenoir Museum: *Springtime* by Marcel Lenoir

Caussade and its Surroundings

Between Montauban and Cahors, Caussade has always taken advantage of its crossroads location. But the town also had to deal with its share of tribulations. It was one of the first victims of the Albigensian crusade since, in June 1209, it was ransomed by crusaders led by the Viscount of Turenne. It was in turn English and French, as tides turned in the battles of the Hundred Years War. A Protestant stronghold, it underwent the painful Catholic reconquest after the bloody episodes of the Wars of Religion. It is no surprise then that Caussade has preserved almost none of its architectural heritage, which was probably quite impressive. Of large urban buildings, only the steeple, in Toulouse style of red brick, and the Arlet Tower from the 13th century have managed to survive the centuries. And near the church, largely rebuilt at the end of the 19th century, the Kaïfa planter's house, with its old-fashioned details, evokes the prosperity that its owner must have enjoyed during the 19th century. A network of *bastides* must have covered the open area of the Aveyron valley. Septfonds, founded in 1271 by Alphonse of Poitiers and, further north, Puylaroque, perched against the sky on the summit of its hill, with its old artisans workshops hidden under Gothic arcades, are still a testimony to the French monarchy's desire to control all these areas through a policy of populating regions bitterly contested by their neighbours, cousins and enemies, the kings of England. Haughty and at the same time fragile on its promontory, the pilgrimage chapel of Notre-Dame-des-Misères is charming with its 16th century steeple, with its gemeled bay openings.

Gothic windows in Puylaroque

Facade of the Kaïfa planter's house in Caussade (detail)

Notre-Dame-des-Misères Chapel

The steeple of Caussade church

Montauban

From the old rose of its brick buildings to the blue-grey of the sparkling waters of the Tarn, many subtle shades blend to paint a harmonious picture of Montauban. Founded in 1144 by Alphonse Jourdain, Count of Toulouse, on one of the last great meanders of the Tarn, the town enjoys an enviable location. Here, in the 14th century, in the heart of a vast alluvial plain, drained by the Garonne, Tarn and Aveyron Rivers, a bridge was built to guard the route into Quercy and Rouergue. Born of this strategic piece of planning, Montauban was, well before its time, one of the first bastides to be built in the southwest.

Faithful to its founding family of noble lords, Montauban wholeheartedly defended their Catharistic ideas. Although never able to take the town by force, Simon de Montfort nevertheless acquired it on the 8th of June, 1215, as homage due from Géraud, Count of Armagnac and of Fezensac, and the terrible judges of the Inquisition came to settle within its walls. As sensitive as ever to theological issues, Montauban adopted Protestantism and became, along with La Rochelle and Nîmes, one of the great Protestant strongholds of the kingdom. Richelieu, however, eventually got the upper hand of Montauban, deeply entrenched behind its fortified walls and its faith, and in 1628, its ramparts were razed to the ground.

During this period of intense spiritual unrest, spurred on by the Counter Reformation, Montauban's old churches were repaired and Notre-Dame Cathedral was built. The bright white stone of this – Our Lady's – cathedral, clearly contrasting with the ochre-pink background of the surrounding buildings, was intended to symbolise the shining victory of Catholicism over the reformed church. The facade is framed by a pair of campanile bell towers and adorned with heavy Doric columns and colossal statues of the four gospel writers. Above the triangular pediment of the narrower

Four Hundred Blows

Louis XIII was nine years old at the death of Henry IV. Marie de Medicis became regent until the young king was freed of it in 1619. From then on, he applied a restrictive interpretation of the 1598 Edict of Nantes. In 1620, he had Catholicism restored in Béarn *manu militari,* in this region where his grandmother, Jeanne d'Albret, had forbidden this religion. Realizing the danger, Protestants held an assembly in La Rochelle, in fact placing the Huguenot community of the French kingdom in a state of alert, but also as outlaws. On April 19, 1621 the King's Council decided on military intervention against "those of the religion". Protestant towns surrendered one after another. Montauban, one of the great Huguenot havens, saw the royal troops arrive at its walls, led by Louis XIII himself and placed under the orders of Constable de Luynes. The siege was put in place on August 16, 1621. It lasted nearly four months. The resistance of the fifteen thousand inhabitants of Montauban, besieged by twenty-five thousand foot soldiers was heroic. Women participated in reinforcing the fortifications of the city, which underwent multiple and intensive cannon attacks. The total number of losses in both camps was about sixteen thousand. On September 18, the Duke of Mayenne perished when he received a musket wound in the left eye. The King believed that he was putting a victorious end to the siege when he ordered "four hundred blows" of the cannon onto a city that was used to it. One day in November, surprised by the calm reigning in the besiegers' camp, the citizens of Montauban found the latter abandoned. Only the wounded and sick remained. The city then had to face another scourge – purpural fever that carried off two thousand of its inhabitants. Eight years later, on August 20, 1629, Richelieu entered a subjected city.

◄ Montauban: Old Bridge, bishop's palace and St. James Church

central part, two triumphant statues represent Faith and Religion. The interior is sober and elegant, and features Ingres' *The Vow of Louis XIII*, painted especially for the cathedral. Saint James' Church, dating from the 13th century, has a high octagonal bell tower that rests on a solidly square tower, surrounded by corbels and machicolations. The portal was completely rebuilt in the 19th century. Above its semicircular recessed arches, a brightly coloured mosaic illustrates *Ezekiel's Vision*.

But the most striking image of the magnificent town of Montauban is that of Pont-Vieux – the Old Bridge – projecting the image of its perfect arches into the bright reflective waters of the river below. Built of red brick, it blends perfectly into a setting of great architectural beauty, at the foot of the former Bishop's palace, which now houses the Ingres Museum. In the background, the bell tower of Saint James' Church keeps watch. Dating from 1311, Pont-Vieux was once flanked by a number of guard towers. Built with a flat deck, which was exceptional in the Middle Ages, and with its piers reinforced by freestone triangular spurs and pierced with fine arched openings to reduce the pressure of water on the structure, the bridge has withstood centuries of capricious fluctuations in its river.

Montauban's old central market square, Place Nationale, still has the same nearly perfectly square shape that was laid out when the town was founded in the 12th century. The original medieval buildings, however, were destroyed by fire and the houses that line the square today date from no earlier than the 17th century. Exceptionally deep covered ways with a double gallery run all the way around the busy square, between shops and market stalls, and the four facades are linked together by stylish porticos, set diagonally at each corner.

Ingres is not Montauban's only great artist and the town boasts many sculptured works by Bourdelle, such as the *1870 War Memorial* in Place Bourdelle, and *The Last Dying Centaur* near the Ingres Museum.

Climbing the Slope

Pierre Paul Riquet, called the "Moses of Languedoc" for having conceived, financed and carried out the *royal canal of communication between the two seas*, since called the Canal du Midi, died exhausted and nearly bankrupt on October 1, 1680. One league was missing for his work to be complete. The canal was filled with water and inaugurated in the following year, joining Toulouse to Sète. Nearly another two centuries were needed for the old dream of linking the Mediterranean Sea and the Atlantic Ocean to be fulfilled. Vauban, who at Riquet's death, had helped in various places with improvements to the canal "to join the two seas", asserted that "he would have preferred the glory of being its author to all that he had done or would do in the future". Vauban had suggested the construction of a lateral canal to the Garonne River, which, with its violent floods and low waters in summer, posed serious navigation problems between Toulouse and the Gironde estuary. A lateral canal came into service only in 1856. To compensate a height difference of thirteen metres, at Montech, south-west of Montauban, a stairway of five locks was created. It was replaced in 1974 by a "water slope", a technique developed by Jean Aubert and used for the first time. The vessel floats in a mobile reach, closed by a mask and moved by two electric railcars on rubber tires along a 3 percent slope that is four hundred and forty-three metres long. The total operation time is not more than six minutes.

Water slope in Montech

Montauban, St. James Church: mosaic showing Ezekiel's vision ▶

The Ingres Museum's Collections

The Ingres Museum was created by the city of Montauban in 1843 in the former bishop's palace overlooking the Tarn River. Jean Auguste Dominique Ingres was born in Montauban in 1780. A pupil of David's, he obtained the Prix de Rome and soon attracted attention by his passionate search for perfection that would make him one of the masters of Academism. An accomplished musician, he played the violin so well that he was the origin of the French expression "*violon d'Ingres*" for a well-chosen pastime in which both virtuosity and passion must be present. As Ingres had wished, the museum exhibits his violin on the sheet music of Mozart's *Don Juan*. The museum's first floor is dedicated to work by Ingres and artists of his time. *Jesus among the Doctors* is a large painting ordered in 1842 by Louis-Philippe. The influence of Raphael and of antique classical art are visible in this work, as well as in *Jesus giving the keys to Saint Peter*. *Roger delivering Angélique* is a small, oval canvas, a copy of the original painting at the Louvre. Angélique, her head tipped backwards in a rather dramatic movement, holds our attention otherwise distracted by the complicated swirls of a knight striking down a writhing dragon. But Ingres was also a master of portraits. That of *Madame Gonse* is a masterpiece of delicacy and exquisite charm. The Ingres Museum contains other marvels, from Bourdelle's vigorous sculptures with his extraordinary *Hercules the Archer* to a fine collection of 18th century ceramics. The very old room of the Black Prince, from the 14th century, contains treasures of medieval art and a horrible torture bench.

Funeral of Lope de Vega by Jean Paul Étienne Balze

Hercules the Archer by Antoine Bourdelle

Portrait of Madame Gonse by Ingres

Moissac

At the foot of the last of Quercy's hills, Saint Peter's Abbey in Moissac stands at the heart of a fertile plain, on the shores of the Tarn River. According to legend, it was founded by Clovis in 506, but it is more likely that, between 628 and 648, Didier, Bishop of Cahors (630-655) had the idea of setting a Benedictine abbey in this place warmed by the sun and with abundant water. Enriched by 680 with donations from Nizezius and Ermentrude, the abbey suffered from the tribulations of history that were to delay its promised fortunes. It was ravaged by Moorish raids in 721 and 732, but rebuilt by Charlemagne (768-814) and his son, Louis the Pious (814-840), devastated once again by the passage of Normans in 850 and 864, then the Magyars in the 10th century. Nearly ruined, the abbey was to disappear when Odilon, the fourth abbot of Cluny (994-1049) agreed that it should become an abbey obedient to Cluny in 1047. This was solemnly ratified by his successor Hugues de Semur (1049-1109) in 1055. Durand de Bredon, a relative and friend of Odilon's, became the first Clunisian abbot of Moissac. Named Bishop of Toulouse in 1059, he raised his abbey to the rank of the most prestigious abbeys of the kingdom. In 1063, he consecrated the new abbey church that he had raised from its ruins, and died in grace at Moissac in 1072. His successor, Hunaud de Gavarret (1072-1085), began work on the cloister and the monastic buildings at the abbey, which became an important staging post on the *via Podiensis*. The abbot from 1085 to 1115, Ansquitil, had the sculptures of the cloister and the abbey church's tympanum made for the arrival of Pope Urban II, who came in May 1097 to consecrate the main altar to the Holy Saviour. While master sculptors displayed their talent in carving Saint John's Apocalypse on the doorway's

The Siege of Moissac

In 1212, in the heart of the Albigensian Crusade, Simon de Montfort undertook the conquest of Lower Quercy. Saint-Antonin-de-Noble-Val was besieged on May 20. The town finally surrendered and its two lords, Adhémar and Pons Jourdain, were imprisoned in Carcassonne. At the request of the bishop of Agen, the head of the crusade then intervened in the Agenais. On his way, he took Caylus and found Montcuq abandoned. Having conquered Penne-d'Agenais, he led his troops to Moissac, where he arrived on August 14. The day before, the city, which had remained loyal to the Count of Toulouse Raymond VI, had received reinforcements of three hundred men, mainly mercenaries. War machines such as catapults and trebuchets were put in place. The siege lasted three weeks. Meanwhile, the neighbouring towns of Castelsarrasin, Verdun-sur-Garonne and Montech submitted to the Lion of the Crusade, although one of the latter's men, the nephew of the archbishop of Reims, was captured by the besieged who killed him, cut him up and catapulted the pieces of his body into the crusaders' camp. Moissac capitulated. Montfort demanded that the mercenaries paid by the Count of Toulouse be handed over and killed them all. He entered the town on September 8 and on the 14, signed an agreement with the abbot of Saint Pierre Abbey in which he took over Raymond VI's role for the town. After the town's surrender, it and the abbey were ransacked by the crusaders. A fire destroyed the upper sections of the cloister and the abbey church as well as the convent buildings. Fortunately, the sculptures of the doorway of the abbey church and cloister were spared. After Moissac, Simon de Montfort decided against attacking Montauban, defended by Roger Bernard, son of the Count of Foix, and decided to turn his attention to the Ariège River valley.

◄ Moissac, pier of south doorway of St. Peter's Abbey Church: Jeremy

tympanum, and on the lintel frieze, in the marvellous monsters of the pier, in Jeremy, Paul or Peter, others – scribes and illuminators – worked in the silence of the scriptorium on the prestigious and unique manuscripts that enriched the abbey's library. The cloister's historiated capitals, living "stone bibles" offered to the view of worshippers, can give an idea of the illuminated treasures of these parchments, today kept at the National Library. Under the Abbot Roger (1115-1131), the doorway and church tower porch were completed, while distinguished relics added to the monastery's fame. Possessing dependencies in Languedoc and as far as Catalonia, the high point of Moissac was reached during the 11th and 12th centuries. The many pilgrims on the road to Compostela could stay at the abbey's hospice, located near Saint Martin's church. Rested and edified, they visited the church of Saint James of Moissac before continuing on their way.

Moissac, cloister of St. Peter's Abbey: a capital

Moissac, tympanum of St. Peter's Abbey Church: old men of the Apocalypse (detail)

East buttress of the doorway: Presentation at the Temple and the Flight into Egypt

Pier of south doorway: Saint Paul

West buttress of the doorway: Feast of the rich man and the Death of poor Lazarus

East buttress of the doorway: Angel Gabriel

Saint Peter of Moissac

The southern doorway of Saint Peter's in Moissac is a masterpiece of Romanesque art in France. In the centre of the tympanum, Christ in majesty is seated, holding the Holy Book in one hand and, with the other, blessing the faithful. Around him, the tetramorph is organized, symbolically representing the four evangelists. Saint Mark, likened to the lion, illustrates the Resurrection since the lion in bestiaries revived his still-born lion cubs with his roars. Saint Matthew, having established the genealogy of Christ and emphasized his humanity, is a man, the image of the Incarnation. The Eagle, symbol of the Ascension, is Saint John, whose thoughts rise towards heaven. Saint Luke, who talks of the sacrifice of the son of man, is represented by the sacrificed ox, an evocation of the Crucifixion. Two seraphim frame the scene. The rest of the tympanum is filled, on three levels, by the twenty-four old men of the Apocalypse, carrying a viola and a bowl filled with perfume, turning and raising their heads towards Christ. Three curves, sculpted with leaves, surround the tympanum that rests on a lintel decorated with three rosaces. On the western side of the monolithic pier, three intersecting couples of lions standing on their hind legs are superposed. On the southern side at the right is the prophet Jeremy, and at left on the north side, Saint Paul. On the piedroits, two bas-reliefs welcome the faithful: on the left Saint Peter, patron of the abbey church, and on the right, the prophet Isaiah. The eastern splay is illustrated with scenes from the life of Jesus: Annunciation, Visitation, Adoration of the Wise Men, Presentation at the Temple, Flight into Egypt. On the western splay, the universe of Damnation is shown in various pictures: the miser and the adulterous woman in the hands of demons, the rich man feasting facing a starving Lazarus. A veritable illustrated book of prayer, this doorway, by its necessarily el-

Moissac, St. Peter's Abbey Church: south tympanum

liptical representations, was intended to edify the faithful who were often illiterate in the Middle Ages. In other words, to teach them the bases of their faith.

Once through the doorway, the visitor is in the narthex with its powerful capitals, such as the one with the wolves fighting over a sheep. A spiral staircase provides access to the upper floor. The nave and chancel, the former re-arranged, the second restored at the end of the 15th century, are Gothic. The nave with its Gothic arches contains four bays. It precedes the chancel, with its three barlong bays forming the transept, and ends with a seven-sided apse. The decoration of the walls, an ochre background on which gray ribbons form little diamonds, is surprising. It is nonetheless a faithful reconstitution of the original decor. In the first southern chapel, to the right of the nave, a finely carved, gilded wooden reliquary from the 18th century contains the relics of Saint Cyprien. The next contains a stone Pietà dated 1476. In the chapel beside it, a fragile painted wooden sculpture, from the 15th century, represents the Flight into Egypt. The last southern chapel of the chancel has another polychrome sculpture from 1485, showing the Laying in the Tomb. In the nave, to the south, a Christ of polychrome wood is contemporary with the tympanum. The cross bearing him, a tree of life with its branches spreading, was made later. On the other side of the nave, an organ case carries the arms of the commendatory abbot of Moissac, Cardinal de Mazarin. Below, in a recess, a Merovingian sarcophagus rests on two Corinthian-style capitals.

The cloister flanks the northern side of the church. It is one of the oldest Romanesque cloisters in France, since the date of its foundation "In the year of the Incarnation of the Eternal Father 1100" is still engraved on the central pillar of the western gallery. It suffered heavy damage during the crusade against the Albigensian heresy, Moissac being besieged by Simon de Montfort in August

Moissac, St. Peter's Abbey: the cloister

1212. But it was restored shortly afterwards. Nearly square, its seventy-six arcades surround a grassy courtyard in which a magnificent cedar tree grows. They are supported alternately by simple and double columns. The columns are topped with capitals. These very flared columns are sculpted with plant, geometric or animal motifs. Several are historiated, repeating themes borrowed from the Gospels and the Old Testament: episodes in the life of Christ, Daniel in the lion pit, or more local themes such as, on one capital in the eastern gallery, the martyrdom of Saint Saturnin, the first bishop of Toulouse. Strong parallelipiped pillars of gray marble mark the four corners and middle of the galleries. They are carved with images of the holy apostles and figures of abbots who marked the history of the abbey, like Durand de Bredons who worked to attach Moissac to Cluny, on the central pillar of the eastern gallery, and Abbot Ansquitil, on the central pillar of the western gallery.

Moissac Museum

Moissac Museum has, since 1951, been housed in the former abbot's palace. Two chapels – one 12th century Romanesque, the other Gothic with large bays – bear witness to the eminent position enjoyed by Moissac's abbots in the past. The museum presents Quercy's heritage, bringing it to life with *in situ* displays such as the 19th century kitchen from lower Quercy where 17th and 18th century dressers and cupboards rub shoulders with fine china from Auvillar. An impressive collection of starched and embroidered headdresses recall the days when delicate openwork and fine lace were the pride of Quercy ladies. The museum pays homage to the popular arts and traditions of the past, as wrought ironwork, pottery and tools, long since fallen from use, paint a portrait of an earlier age.

Moissac, capital in St. Peter's Abbey cloister: Daniel in the lion pit

Bastides and Castles in White Quercy

The rolling landscapes of White Quercy develop between Cahors in the north-east and Moissac in the south. The chalky substratum of these hills sometimes shows on the surface and has given its colour to this southernmost part of Quercy, which extends from the Lot to the Tarn Rivers. The bastide of Castelsagrat was founded in 1270 by Alphonse of Poitiers. The Church of the Assumption possesses a large, 17th century reredos. In Baroque style, it is decorated in a very detailed manner with twisted columns, enlaced by supple vines inhabited by birds and squirrels. The grapevines that cover the extremely sunny hillsides in the region produce the famous Chasselas grapes of Moissac. Lauzerte quarrels over the slopes of a hill with vineyards and orchards. This bastide, founded in 1241 by Raymond VII of Toulouse, has maintained its covered square and half-timbered houses. Saint-Barthélémy Church, with its large rosace, is in Gothic style. Montcuq received its customary charter from Raymond VI of Toulouse. It has retained only the 12th century keep from its fortified castle. In the interlacing of its sloping streets, its half-timbered houses and Gothic arcades huddle around Saint-Hilaire Church. Castelnau-Montratier is a village on a crest that overlooks the two parallel valleys of the Barguelonne to the north and the Lutte to the south. In the Middle Ages, the town of Castelnau-de-Vaux occupied the bottom of the little valley and the lower slopes of the promontory on which the fortress had been built. During the crusade against the Albigensian heresy, the lord of Castelnau had remained loyal to the Occitanian cause against the troops of Simon de Montfort who, in 1214, took the town and sacked it.

Castelnau-Montratier church

Several years later, the lord Ratier de Castelnau, instead of raising the town up from its ruins, preferred to create a well fortified bastide on the hill. Castelnau-Montratier was born. Its triangular central place is surrounded by arcades. The shade under the thick foliage of its lime and chestnut trees is welcoming in the summer. The town hall is topped with a belfry whose shape is softened by the curving lines of its beams. The church of Castelnau-Montratier, built at the top of the promontory, overlooks the village and the surrounding valleys. With its rounded dome, its towers topped with small spires supported by little colonnades and light walls, it resembles Sacré-Cœur Church in Paris. A few kilometres away, on the road to Cahors, Boisse windmill is still in working condition.

Town hall belfry,
Castelnau-Montratier

Montcuq keep

Windmill, Boisse

Marble recumbent statue of
Pierre des Près

Facade of St. Martin's
Collegiate Church

Flemish tapestry showing
Saint Martin

St. Martin's Collegiate Church,
Montpezat-de-Quercy

Montpezat-de-Quercy

Montpezat-de-Quercy is a little perched town that has retained its medieval appearance. A gate made of white causse stone opens onto the charming picture of its narrow streets. Corbelled houses and Gothic arcades have been well restored. In the heart of the town, a square surrounded by ancient residences has a covered walkway supported by heavy stone pillars. The town received its franchise from the Count of Toulouse in 1257 and was pillaged by the English during the Hundred Years War. But the little medieval city rapidly grew wealthy and beautiful through the impetus of the rich des Prés family, who endowed it with a collegiate church in which a fabulous treasure accumulated.

Pierre des Prés, the first in a long series of prelates, all of whom came from the same family clan, was a prominent cardinal in Rome. His nephew, Jean, became the bishop of Castres and of Coimbra in Portugal. Three bishops of Montauban also belonged to this family. The last of them, Jacques des Prés, was named a bishop right in the middle of the Wars of Religion. Weapons in hand, he defended the prerogatives of his faith in the heart of his bishopric, which was strongly influenced by the new reformed religion, Protestantism. Pitched battles, guerilla warfare, ambushes, and plots multiplied during this holy war. In 1589, the soldier-bishop of Montauban came to a tragic end, the victim of an ambush in Lalbenque, about fifteen kilometres from Montpezat-de-Quercy. So there is no surprise that the little collegiate church of Saint Martin in Montpezat-de-Quercy should become one of the best supplied in religious furnishings in the region. It was built during the 14th century according to the simple plan of single-nave churches with side chapels found all over Languedoc. Its completely disssymmetrical facade is flanked on one side by the square mass of a large tower, and on the other, by a simple bevelled buttress. A rose-window opening above the doorway allies grace and elegance in its simplicity. The long nave has Gothic arches with the arms of the des Prés family on the keystones. The marble recumbent statue of Cardinal Pierre des Prés, who died in 1366, is matched on the other side of the chancel by the stone statue of Bishop Jean des Prés, dead in 1351. Fine statues of the Virgin, such as the slender, alabaster Virgin with Doves from the 14th century, or the polychrome sandstone Virgin from the 15th century, but also reliquaries made of wood, silver or rock crystal, the altar, the stalls and the baptismal fonts are all only part of the extraordinary treasures that have accumulated in this little collegiate church.

In fact, Saint Martin's church contains a series of five Flemish tapestries that give its religious furnishings matchless value. Manufactured in the 16th century, they were designed for the decoration of this church and have adapted to the surrounding decor with great felicity. These monumental tapestries, twenty-five metres long and two metres wide, often weighing about a hundred kilograms per square metre, they are nevertheless masterpieces of refinement and detail.

In sixteen pictures, they recount the exemplary life of Saint Martin. On a dark red banderole running along the top of each scene, an octosyllabic quatrain of Gothic letters is embroidered in French. Saint Martin, wearing a purple coat is mounted on a high-spirited steed. But a poor beggar, a lame leper, limps toward him and looks enviously at the heavy cloth hanging from Saint Martin's shoulders. The saint has taken out his sword and prepares to divide up his coat. The picture of the Devil is particularly well done, with his pustular, pink and hairy skin, his long tail, his clawed feet, his long ears and his tongue hanging out. Saint Martin's victorious struggle against Evil is shown in a simple, but highly efficient, glorification.

INDEX

TO

WILLIAM

LOVE FROM

....

William loves being under the sea. Whether it's exploring a sunken pirate ship, swimming with mighty whales or looking at big shells, he couldn't be happier.

But today was different. William was about to meet the most majestic creature of the underwater world....

William was gliding through the cool, clear water, hypnotised by the coral, when he heard something strange.

"Is that ... crying?" he said.

SOB! SOB! SOB!

William swam towards the sound and stopped when he spotted an entrance to a deep, dark cave....

"I wonder what's in there?" William thought as he drifted closer.

As he bobbed up and down a sea creature peeked out from the darkness!

"Err, he- hello?" he whispered.

William could just about make out that he was pink with blue sticky-up hair. But he wasn't scary, he just looked very sad.

And he was carrying a gold crown.

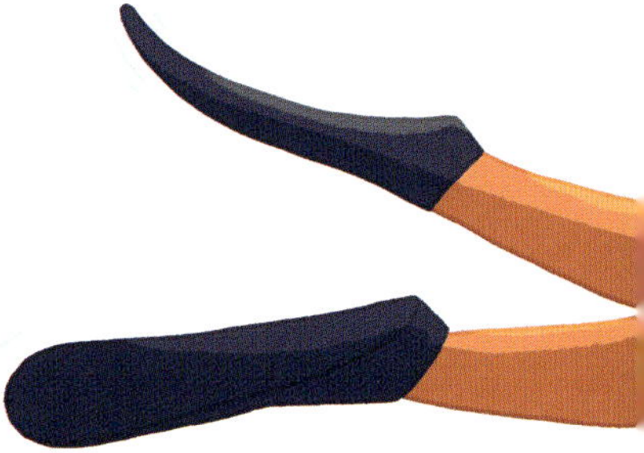

The sea creature wasn't just any old sea creature – he was a king. But he might not be for much longer.

"I was having my afternoon snooze, and when I woke up the jewels from my crown were gone," he sniffled. "If I don't get them back, how can I be king?"

"Don't worry," said William. "I'll help you find them. Whoever took them can't have gone far. Let's go!"

And so the search for the lost jewels began....

"There's one!" pointed William.

"Oh no, the grumpy crabs have got it!" gasped the king.

SNAP! GRUMBLE, GRUMBLE! SNAP! SNAP!

William quickly dodged round one, darted past another, ducked under a third and rescued the jewel from a snapping claw.

"I've got it!" he yelled.

"Quick, let's skedaddle," said the king. "They'll be grumpier than ever now!"

Up ahead, William and the king saw a giant clamshell perched on a rock. Could a jewel be trapped inside?

"Hmm," said William, "how do you get a clamshell to open up?"

"I don't know," said the king, "how do you get a clamshell to open up?"

"It's not a joke!" said William, "but that gives me an idea...."

He grabbed a handful of seaweed and began to tickle.

HA HA HA! HA HA HA! HA HA HA!

The clamshell burst open and the king plucked out the jewel.

"Ta-dah, only two more left to find!" he cheered.

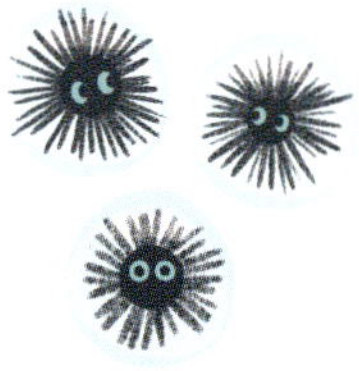

Suddenly, a dolphin whooshed past and darted upwards.

"Something fishy's going on," giggled William, "let's go!"

They burst onto the surface where a pod of dolphins was jumping and spinning and diving.

SPLASH! SPLISH! SPLOSH!

And they were throwing around another jewel.

William had to act fast. He threw his wallet up in the air to trick the dolphins. Then, he quickly caught the jewel.

"Three down, only one to go!" he smiled.

The only place left on the reef they hadn't searched was ...

THE SEAWEED FOREST!

"Are you s-s-sure you want to g-g-go in there?" stuttered the king nervously.

William looked up at the towering hairy branches swaying back and forth. Big, spooky eyes blinked back at him from the darkness, but that didn't stop him.

"I'll be fine!" said William as he flicked on his torch.

William hadn't been in an underwater forest before, but it was fun.

As he swam upwards he suddenly came face-to-face with the cutest thing he'd ever seen – a gigantic octopus.

The octopus looked startled and darted off. And as he did, eagle-eyed William spotted the last jewel.

"Come back!" he shouted.

But the octopus disappeared.

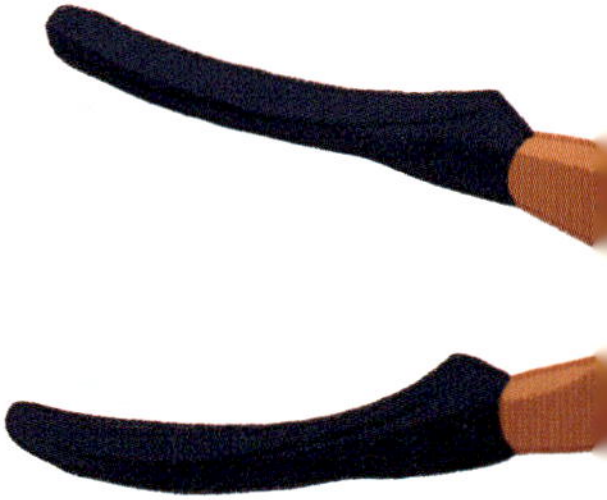

William needed a plan.

He kicked his flippers harder and harder and swam faster and faster through the seaweed to get ahead of the octopus.

Then he stopped, hid and kept very still.

It wasn't long before the seaweed started to rustle. He waited for just the right moment, and then....

"BOOOO!" yelled William.

"AHHHHHHHH!" wailed the octopus.

William quickly jumped out and rescued the final jewel.

William carefully pressed the jewels into the king's crown and swam up to place it on his head.

"Ready?" asked William.

"Ready!" smiled the king.

As William lowered the crown all the fishy friends let out a big cheer, and all around them looked like awesome fireworks.

WOO-HOO! HOORAY!

"William, you're my hero!" declared the king.

"I can't thank you enough," said the king. "If there's anything I can do for you, just name it."

"Well, there is one small thing," smiled William. "Can you please let the jewel thieves out? They were only playing."

"Consider it done," said the king, "I guess they've learned their lesson now."

And with that, William turned, waved goodbye and began to swim home, daydreaming of his next BIG adventure....

TIME OUT

MEET SANTA?
Maybe I'll...

William, look out for more **Mini Adventures** books.
Go to www.orangutanbooks.co.uk

Story by Jane Kent
Illustrated by Jo Lindley
Designed by Jane Gollner

orangutan

First published by Orangutan Books in 2019
Hometown World Ltd, 1 Queen Street, Bath BA1 1HE

Visit
www.orangutanbooks.co.uk
Follow us @orangutanbooks

ISBN 978-1-78979-504-2

Printed in Italy
HTW_PO201907

FSC
www.fsc.org
MIX
Paper from
responsible sources
FSC® C023419